The Health Benefits of Omega-3 Fatty Acids

In

Inflammatory Bowel Disease

And

Irritable Bowel Syndrome

By

Maria Martin

Copyright © 2012
Maria Martin
All rights reserved

No part of this book may be reproduced or transmitted in any form or by any means, electronic or mechanical, including photocopying, recording, or by any information storage and retrieval system, without the written permission of the author and publisher, except where permitted by law.

Published by Lulu Press, Inc.

ISBN 978-1-300-01784-4

Printed and published in the United States of America

Table of Contents

Preface..5

Abstract..9

Introduction...13

Structure of the Digestive System..17

Proteins and Carbohydrates..25

Fats: The Good and the Bad...29

Development of Problems in the Gastrointestinal Tract..................................35

Irritable Bowel Syndrome..39

Inflammatory Bowel Disease...43

Leaky Gut Syndrome..49

Essential Fatty Acids..53

Eicosanoids..63

The Role of Cytokines in IBD..69

Anti-inflammatory Effects of Omega-3 Fatty Acids in IBD..............................73

Serotonin and IBS...79

Imbalance of Omega-3 to Omega-6 Fatty Acids - 85
Effects of a Modern Day Diet

Current Medical Treatment for IBS and IBD...89

World Demographics Comparing Rates of IBS and IBD.................................95

Conflicting Views and Medical Studies..101

Conclusion..105

Bibliography..111

Preface

This book is an updated version of a thesis that was written and submitted to Clayton College of Natural Health in 2004 in order to receive my Master's degree in Holistic Nutrition. It was recommended at the time that I publish the paper, but I failed to do so because of time constraints. Things have slowed down for me recently, so I decided to go ahead and look into publishing the paper.

Before I decided to publish this book, I was somewhat concerned that this paper was no longer relevant since I wrote it eight years ago. However, after re-reading it, I realized that it is, in fact, still quite relevant in today's world. The value and benefit of fish oil and omega-3 fatty acids is being recognized more and more with each passing year. These essential fatty acids are being touted as extremely beneficial in all kinds of medical diseases, especially cardiovascular disease. This book goes into detail about the structure and function of these amazing fatty acids and the beneficial role that I believe they play in irritable bowel syndrome and inflammatory bowel disease. I'm sure with time we will hear even more about the benefits of omega-3 fatty acids in other diseases.

At the time I wrote this paper, I had been given the diagnosis of irritable bowel syndrome. As it turns out, I actually had a condition called adenomyosis, a uterine disorder that causes severe pain and heavy bleeding. I found that omega-3 fatty acids helped me tremendously during this time. Although I was ultimately diagnosed with adenomyosis, both conditions involve inflammation, and omega-3's have been shown repeatedly to help in inflammatory conditions. This just goes to show the tremendous benefit that this little fatty acid has on the human body and in many different medical conditions.

I hope you enjoy this book as much as I had writing it. I learned a great deal and hope that you too will realize the tremendous impact that omega-3 fatty acids can have on a person's health!

Abstract

This book explores the use of omega-3 fatty acids in the treatment of inflammatory bowel disease and irritable bowel syndrome. The role of fats is discussed and clarified as some fats are vital to health. Both inflammatory bowel disease and irritable bowel syndrome are described in detail with discussions of symptoms, factors that may exacerbate the conditions and dietary advice for sufferers. Leaky gut syndrome, seen in both disorders, is presented as a possible factor in the development of inflammation in the gastrointestinal tract. Since omega-3's are precursors to anti-inflammatory substances in the body, the functions of these fatty acids are examined in detail and their role in the reduction of inflammation is discussed. In addition, the role of omega-3 fatty acids in the regulation of the neurotransmitter serotonin is addressed since a malfunction of the serotonin signaling system has been observed in irritable bowel syndrome. Omega-6 fatty acids are also discussed for their role in promoting inflammation. The ratio of omega-6 to omega-3 is compared between today's diet and that of our ancestors, and the differences are shown in relation to the dramatic modern day increase in inflammatory conditions such as inflammatory bowel disease. Many different studies are presented that suggest omega-3's, with their anti-inflammatory and serotonin regulating properties, may play a prominent role in the treatment of these disorders. Current medical treatments are detailed along with a discussion of their side effects for both disorders. The consumption of omega-3 and omega-6 fatty acids and their corresponding rates of inflammatory bowel disease are reviewed along with information on irritable bowel syndrome. In conclusion, naturally occurring omega-3 fatty acids offer an alternative to medicine without side effects and, along with a reduction in the intake of omega-6's, may be an effective form of treatment for these disorders.

# *Introduction*

"One third to one-half of all adults have digestive illness - over 62 million people."[1]

In today's society, negative connotations are associated with the word "fat" and diet. The low fat craze has led to chronic health problems which have in turn promoted a closer look at fats in the modern day diet. The understanding of the role of fat in the diet is misunderstood as some fat is necessary and essential for the proper functioning of the human body. Fats are broken down and stored in the body in the form of triglycerides which contain molecules called essential fatty acids, two of which are omega-3 and omega-6.

Crohn's disease (CD), ulcerative colitis (UC) and irritable bowel syndrome (IBS) have become very common in recent times and are seen with high frequency in people who consume a "westernized" diet. Diets high in processed foods and fat have high amounts of omega-6 and very few omega-3 fatty acids. These factors may play a big role in the increase in frequency of these inflammatory conditions. Although essential in the body, omega-6 fatty acids promote swelling and inflammation which is a predominant feature in Crohn's disease and ulcerative colitis, collectively referred to as inflammatory bowel disease or IBD. As stated in Alternative Medicine: The Definitive Guide by Burton Goldberg, M.D., "The typical Western diet of high-fat, high-carbohydrate, highly processed foods with many additives and preservatives is the root cause of many digestive disorders."[2] There is also strong evidence of a lack of omega-3 fatty acids in the American diet. In addition to their anti-inflammatory properties, omega-3 fatty acids have been shown to regulate serotonin, a neurotransmitter found predominately in the digestive tract. Malfunctioning of the serotonin system in the gastrointestinal tract has recently been shown to play a role in IBS.

My interest in this subject has come from personal experience. I was diagnosed with irritable bowel syndrome about 7 years ago. While under the care of a family physician, a reproductive endocrinologist and a gastroenterologist, I underwent a colonoscopy and laparoscopy which did

---

[1] Elizabeth Lipski, <u>Digestive Wellness</u> (Los Angeles: Keats Publishing, 2003) 3.
[2] Burton Goldberg, <u>Alternative Medicine: The Definitive Guide</u> (Tiburon: Future Medicine Publishing, Inc., 1999) 680.

not prove to be helpful. I was put on many different medications in an effort to control the pain, none of which worked. Finally relief came when I started to sprinkle flaxseed meal, a source of omega-3 fatty acids, on my food. I had read about the benefits of the omega-3 fatty acids but did not expect to get relief from IBS. My symptoms dropped dramatically, and with continued changes to my diet, I was able to overcome this disorder. I have not had a full blown attack for six years.

Medications are currently available for the management of both IBS and IBD. These medications work to reduce pain and inflammation in IBD and help regulate serotonin in IBS. Although effective, these medications also have some serious side effects. Both personal experience and research have suggested that omega-3 fatty acids may be able to help these disorders without the side effects of medication. Also, reducing the intake of omega-6 fatty acids by making dietary changes can significantly affect the progress of these disorders by balancing out the ratio of omega-6 to omega-3 fatty acids. Because of the anti-inflammatory and serotonin enhancing effects of omega-3 fatty acids, it is likely that this supplement can play a major role in decreasing the symptoms of both inflammatory bowel disease and irritable bowel syndrome.

Structure of the Digestive System

To see how omega-3 fatty acids work in the digestive tract, it is necessary to understand the anatomy of the digestive system. It is especially important to understand how peristalsis is controlled since its malfunction is the main problem seen in IBS. The role of the neurotransmitter serotonin is important as it controls peristalsis, and abnormal amounts of this neurotransmitter in the digestive tract have been observed in IBS.

The gastrointestinal tract is 25 to 32 feet long and runs from the mouth to the anus. The parts of the tract include the mouth, pharynx, esophagus, stomach, duodenum, jejunum, ileum, cecum, ascending colon, transverse colon, descending colon, rectum and anus.

The first stage of digestion is chewing which is extremely important since improper chewing can lead to other health issues. If food is not properly chewed, an increase in hydrochloric acid (HCl) production in the stomach can occur leading to ulcers. Also, large food particles that make it to the bloodstream can set off an allergic reaction in the body. Teeth and smooth muscle in the mouth contribute to the mechanical operation of chewing, and saliva begins to digest carbohydrates through the action of the digestive enzyme amylase. Chewing also stimulates the parotid glands which results in a release of hormones that cause the thymus gland to produce T-cells, an important part of the immune system.

The esophagus transports the food from the mouth to the stomach by peristalsis. Food that has been chewed well will pass to the stomach in about six seconds; however, dry food that has not been chewed well may take as long as several minutes. At the bottom of the esophagus is the esophageal sphincter which only opens when food is swallowed and entering the stomach. Otherwise it remains closed and prevents stomach acid and food from coming back up into the esophagus. As food enters the stomach, it is further digested by the action of HCl and pepsin which liquefies the food into a substance called chyme. HCl is produced by the parietal cells in the stomach lining and works to break down proteins. Pepsin further breaks down proteins into smaller groups of amino acids. The stomach also produces the enzyme lipase which helps to break down fats.

Mucopolysaccharides and prostaglandins protect the stomach lining from the acid. Low amounts of these protective substances can lead to ulcers due to the action of HCl on the stomach lining. Low HCl levels due to declining production from parietal cells produces inefficient digestion, and this poses a problem as we age. Food stays in the stomach from two to four hours and then passes into the first part of the small intestine, or duodenum, through the pyloric valve.

Poor digestion is especially important concerning the absorption of vitamin B12. A substance called intrinsic factor is also produced by the parietal cells of the stomach lining. This substance joins with extrinsic factor, or vitamin B12, in the stomach. Intrinsic factor helps vitamin B12 to be absorbed in the intestines, so a lack of intrinsic factor can lead to a deficiency of vitamin B12.

The liver, pancreas and gallbladder contribute digestive juices necessary for the further breakdown of food. The pancreas releases alkaline enzymes which reduce the acidity of the stomach acid. Three enzymes released by the pancreas are protease, amylase and lipase. Protease helps to break down proteins; amylase breaks down carbohydrates; and lipase helps to digest fats. The liver detoxifies toxins, produces cholesterol, stores the fat soluble vitamins A, D, E, and K and produces bile. Bile released by the gallbladder is alkaline and partly responsible for fat absorption. It is also active in the breakdown of cholesterol.

The small intestine, also called the small bowel, is twenty feet long and has three parts: the duodenum, jejunum and ileum. Nutrients are absorbed into the bloodstream through microvilli and the lining of the tract replaces itself every three to five days. The duodenum is only a few inches long and is responsible for the absorption of iron, calcium, copper, folic acid, zinc, thiamine, manganese and vitamin A. The jejunum and ileum are each about 10 feet long, and most of the nutrients from food are absorbed into the bloodstream in the jejunum. Fats are absorbed more slowly and require both the jejunum and ileum for absorption. Vitamin B12 and bile salts are absorbed exclusively through the ileum.

The ileocecal valve is a valve between the small and large intestine near the cecum and appendix. It helps to regulate movement of material through the intestinal tract. If it remains open too long, diarrhea will result, and constipation occurs if it remains closed for too long. Stimulation of the sympathetic nervous system increases the contraction of the ileocecal valve. The appendix, once thought to have no function, is now believed to be part of the immune system due to the presence of large amounts of lymphatic tissue. This tissue produces lymphocytes which are important white blood cells used to fight off infection.

The large intestine or colon is about five feet long and begins at the cecum. Some final products of digestion, electrolytes, and water are absorbed in this part of the gastrointestinal tract. The colon consists of three longitudinal bands, called teniae coli, that run down the middle of the colon. The teniae coli are shorter than the rest of the muscles in the colon. As a result, pouches, called haustra are formed, and these pouches aid in water absorption by slowing the passage of food residue. The ascending colon proceeds from the cecum and leads to a bend called the hepatatic flexure. This leads to the transverse colon until it reaches another bend called the splenic flexure. From here, the descending colon moves down the left side until it reaches the sigmoid colon which proceeds to the rectum and then to the anus.

As food distends the stomach, the gastrocolic reflex is initiated causing the contraction of the rectum. The sympathetic nervous system plays an excitatory role and the parasympathetic nervous system plays an inhibitory role in the operation of the internal anal sphincter which controls defecation. The external sphincter is voluntary. If the urge to defecate is ignored, the colon will continue to absorb water which leads to dry stool and constipation. The contents of stool are 75% water, 8% live and dead bacteria, 4% undigested food (fiber), 3% dead tissue, 1% mucous with the remaining 9% consisting of a variety of different substances.

There are four layers that make up the wall of the gastrointestinal tract: the mucosa, the submucosa, the muscularis and the serosa. The mucosa is the innermost layer and contains enteroendocrine cells which secrete hormones in the GI tract. Some of these cells, called enterochromaffin cells, secrete serotonin. Motilin, a 22 amino acid residue, is also secreted by these cells and induces contraction of smooth muscle. The submucosa is composed of smooth muscle fibers. The muscularis contains two layers of smooth muscle: an outer longitudinal layer and an inner circular layer. The serosa is the outermost layer of the wall.

Although the GI tract is partially regulated by the autonomic nervous system, it is predominantly regulated by its own nervous system separate from the autonomic nervous system. This is called the enteric or intrinsic nervous system. This system contains about 100 million neurons which are about the same number that is found in the entire spinal cord. The enteric nervous system is connected to the central nervous system through sympathetic and parasympathetic fibers. The parasympathetic nervous system increases the activity of smooth muscle in the GI tract, and the sympathetic nervous system decreases activity.

Between the outer longitudinal and inner circular layers of the muscularis is a group of neurons called the myenteric or Auerbach's plexus. Another plexus, called the submucous plexus, is located between the inner circular muscle of the muscularis and the mucosa. The myenteric plexus in concerned with motor control of the GI tract, while the submucous plexus is concerned with control of intestinal secretions, regulating blood flow through the GI tract and controlling functions of the epithelial cells. There are also some minor plexi present beneath the serosa and in the mucosa.

Within each plexus, there are 3 types of neurons: sensory, motor, and interneurons. Sensory receptors can sense the state of the gastrointestinal wall and the contents of the GI tract. A type of sensory neuron called a chemoreceptor can sense what substances are present, such as glucose and amino acids, and can sense the amount of tension present in the intestinal wall. The motor neurons control secretion, absorption and motility while the interneurons relay the information back and forth between the sensory and motor neurons. The importance of these plexi and their neurons can be seen in a congenital abnormality of the digestive system called megacolon. In this disorder, bowel movements are infrequent (once every couple of weeks) and is due to the absence of ganglion cells in the myenteric and submucosa plexi in the distal colon.

Peristalsis is the process by which food is moved through the gastrointestinal tract. It occurs from the esophagus to the rectum. While there is contraction behind the stimulus, or bolus, there is at the same time an area of relaxation in front. This activity propels food at a rate of 2 to 25 cm/s. The rate of this process can be increased or decreased by the autonomic nervous system. The specific action of this mechanism involves the release of serotonin after the lumen of the intestine is stretched. Serotonin activates the myenteric plexus which in turn stimulates neurons to release acetylcholine and substance P. Substance P increases the motility of the small intestine, but both substances cause smooth muscle contraction. At the same time, other neurons stimulate the release of the gas nitric oxide (NO), the hormone vasoactive intestinal polypeptide (VIP) and the purine adenosine triphosphate (ATP), all of which produce relaxation ahead of the stimulus. VIP also stimulates the intestinal secretion of electrolytes which increases the amount of water in the GI tract. It also dilates peripheral blood vessels and inhibits gastric acid secretion. This substance is also found in the brain and consists of 28 amino acid residues.

As you can see, the gastrointestinal tract is very complex and requires the proper functioning of many different systems for efficient digestion. Digestive enzymes and secretions are important in digestion; however, many other factors such as hormones, neurons and neurotransmitters are also vitally important in the functioning of the GI tract. Proper transport of food through the system can go terribly wrong when these vital substances become imbalanced. This has been shown in IBS when serotonin functioning becomes abnormal.

Proteins, Carbohydrates and Fiber

Foods contain proteins, carbohydrates, fats and/or fiber. Fats will be discussed in depth later. Carbon, hydrogen and oxygen are obtained from the intake of any of these types of foods, but nitrogen can only be obtained from protein. These elements are necessary for the functioning of the body at the cellular level.

Proteins are important in the development of skin, teeth, and bones. In addition, they build and repair damaged tissue and are important in enzyme and hormone production. Proteins are broken down by gastric juices and are difficult to digest since they use more energy in the digestion process than do other types of food. Deep frying and charbroiling make them even more difficult to digest. The typical American eats two to ten times the amount of protein than they actually need. Too much protein in the diet puts a strain on the kidneys resulting in damaged renal tubules and decreased kidney function. It also forces the kidneys to excrete excess calcium and can lead to digestive disorders.

Proteins can come in two forms: complete and incomplete. Complete proteins are those that contain all eight essential amino acids and include tryptophan (a precursor to serotonin), lysine, methionine, phenylalanine, threonine, valine, leucine and isoleucine. These are present in animal products such as poultry, beef, pork, milk, and eggs. Incomplete proteins are those that do not have all eight essential amino acids present, but can be combined to form a complete protein. Plant foods are sources of incomplete proteins, and an example of food combining which will result in a complete protein is rice and beans.

Carbohydrates supply us with a steady supply of energy throughout the day. Complex carbohydrates are better than the simple carbohydrates and are a rich source of fiber, vitamins, and minerals. Carbohydrates are broken down quickly into mono-, di-, or polysaccharides, and energy is released rapidly. Monosaccharides are simple sugars that are absorbed by the body without changing form. Polysaccharides are complex sugars broken down partially by the enzymes in saliva. The digestion is completed in the small intestine where they pass into the blood and are used by cells for energy. Any unused energy is stored in the liver as glycogen. Some examples of carbohydrates include whole grains, raw fruit, legumes, vegetables, nuts, and seeds.

Dietary fiber is indigestible material that reaches the large intestine unchanged. Fiber stimulates peristalsis, absorbs excess water, and combines with residue in the large intestine to create bulk which is excreted. Although fiber contains no nutrients, it is important in the cleansing and detoxifying process of the colon.

Protein, carbohydrates, and fiber play a vital role in maintaining health. Although the focus of this book is on the vital role of healthy fats in the maintenance of IBD and IBS, the components of a healthy diet should include a proper balance of proteins, carbohydrates, healthy fats, and fiber.

Fats: The Good and Bad

"Fat has become a foul three-letter word in our society. We've become a nation of fat phobics, and some of us try to avoid this nutrient at all costs in an effort to lose weight and improve our health. Yet this war on fat has been completely misguided."[3]

Fat is needed for both immediate and reserve energy since it supplies two times as many calories per molecule as compared to carbohydrates or protein. It is important in the formation of enzymes, hormones, cell walls, and brain tissue. Greater that sixty percent of the dry weight of the brain is fat. Fats are also needed to help absorb the fat soluble vitamins A, D, E, and K, and they help to cushion organs and to keep the body warm. In addition, fats are required for nerve insulation and aid in the proper functioning of nerve synapses.

Fats, also referred to as lipids, are made from hydrogen, carbon, and oxygen and are insoluble in water. Of the lipids in the human body, ninety five percent are triglycerides with the other five percent consisting of phospholipids such as lecithin and sterols such as cholesterol. Fatty acids, which are stored in the body in the form of triglycerides, consist of a carbon chain with a carboxyl group (COOH) attached at the end of the chain. The carbon chain is hydrophobic (repels water) while the carboxyl group is hydrophilic (attracts water). These fatty acids are usually twelve to twenty four carbons in length. The functions of the essential fatty acids include regulation of the metabolism of cholesterol and maintenance of the integrity of cell membranes. They are also precursors to prostaglandins, thromboxanes, and prostacyclins which are three vitally important hormones that play significant roles in the inflammatory process.

Fats can be divided into two types: saturated and unsaturated. In a saturated fat, every carbon atom is linked to a hydrogen atom. No more hydrogen atoms can be accepted by the molecule as illustrated below:

```
    H   H   H   H   H
    |   |   |   |   |
H---C---C---C---C---C---H
    |   |   |   |   |
    H   H   H   H   H
```

---

[3] Barry Sears, <u>The Omega Rx Zone: The Miracle of the New High-Dose Fish Oil</u> (New York: Harper Collins Publishers, Inc, 2002) 20.

The previous example shows that all carbons are linked to the maximum number of hydrogen atoms possible. These fats are solid at room temperature and will increase cholesterol. Examples are butter, meat and lard.

Unsaturated fats are not completely saturated with hydrogen atoms. Since not all carbon atoms are paired with a hydrogen atom, double bonds form between carbons. The more double bonds a fat has, the more unsaturated it is. An unsaturated fat is illustrated below:

```
        H                   H                   H
        |                   |                   |
    H---C---H               C               H---C---H
         \                / | \                /
          C=====C    H   C=====C
          |     |       |     |
          H     H       H     H
```

The carbons with the double bonds can accept another hydrogen atom and are therefore unstable and reactive. Also note that the chain is now kinked as opposed to the saturated fat where the chain is straight. The more "kinks" that are present, which is directly related to the number of double bonds between carbons, the more fluid the fat. Since polyunsaturated fats have several missing hydrogens and several double bonds between carbons, these fats remain liquid either at room temperature or in the refrigerator. These oils have been shown to help prevent heart disease. Examples are fish oil, corn oil, soybean oil and sunflower oil. Two very important polyunsaturated fats are the long chained omega-3 and omega-6 fatty acids which are discussed later.

A monounsaturated fat has only two missing hydrogen atoms and therefore only one double bond. They are liquid at room temperature but cloudy in the refrigerator and have been shown to decrease cholesterol. Examples are olive, peanut and canola oil.

There are three categories of lipids: simple, compound and derived. Simple lipids consist of the mono-, di-, and triglycerides. The mono- and diglycerides help to make hydrogenated products more pliable. The triglycerides are the major storage form of fatty acids. They consist of one glycerol molecule and three fatty acid molecules. Compound lipids contain a non-lipid part in their composition. Lecithin, glycolipids, and lipoproteins are examples of compound lipids. Lipoproteins are surrounded by a protein coating which reacts with water. Therefore, it can be easily transported in the bloodstream. Derived lipids are fatty acids that are related in some way to co-enzyme A. One example is cholesterol.

Fats can be found either in their natural or unnatural forms. Cis fats are in their natural form and are the healthiest way to consume fats. The hydrogens of the fatty acid are on the same side of the carbon chain. Since these hydrogens naturally repel each other, the carbon chain will bend away from the hydrogen side. These kinks help the cell to be more fluid and flexible which allows a healthy exchange of nutrients in and out of the cell. Trans fats are the unnatural form of fats and are extremely unhealthy. Since the hydrogen atoms are on opposite sides of the carbon chain, no bending of the chain takes place and it remains straight. To make trans fats, a hydrogen atom is added to an unsaturated fat through a process called hydrogenation. This procedure is used to prolong the shelf life of the product since unsaturated fats tend to oxidize and turn rancid easily.

Hydrogenated fats are implicated in inflammatory conditions, cancer and heart disease. Some examples of a trans fat are partially hydrogenated vegetable oil, solid shortenings, hydrogenated lard, and solid margarines. These fats contribute to high cholesterol and clogged arteries by infiltrating cell membranes and reacting with enzymes so that fatty acids can't do their job. According to Elizabeth Lipski in her book Digestive Wellness, "European research has shown that essential fatty acids found in the cis formation are necessary for electrical and energy exchanges that involve sulfur-containing proteins, oxygen and light. Trans fatty acids are not suitable in these processes and jam the "plug" for the cis fats. These electrical currents are responsible for all body functions, from the way our minds work to heartbeat, cell division, muscle coordination, and energy levels."[4] These dangerous fats can be found in the spleen, liver, muscles, heart, adrenals, and breast milk. They obstruct twenty to forty percent of enzymes and increase the body's need for essential fatty acids. Six to eight percent of the total calories in the average American diet consists of trans fats.[5] In addition, the process of hydrogenation destroys omega-3's leaving

---

[4] Lipski, 159.
[5] Lipski, 160.

those who consume a highly processed diet dangerously deficient in these essential fatty acids.

The misguided concept that all fat is bad for your health is incorrect and dangerous. Although some fats such as saturated, hydrogenated and trans fats are extremely unhealthy and can cause a myriad of health problems, monounsaturated and polyunsaturated fats are not only healthy but essential for normal body functions.

Development of Problems in the Gastrointestinal Tract

"Many physicians believe that the underlying cause of digestive illness is a combination of poor nutrition and exposure to toxic substances."[6]

Many factors play a role in the development of gastrointestinal disorders. Some common causes of these disorders include the following:

> Poor diet
> Food allergies
> Stress
> Infections
> Too high or too low hydrochloric acid levels
> Insufficient exercise

Although we will focus on the effects of diet on digestive disorders, all of the above factors play a critical role in the optimal functioning of the digestive tract. Each of the above factors needs to be addressed in the treatment of any digestive disorder.

A poor diet can have dramatic health consequences. Processed foods are one of the main factors leading to health problems because they are lacking in the enzymes needed for proper digestion. In addition to interfering with liver detoxification, trans-fatty acids inhibit enzymes that are used in the synthesis of fatty acids and cholesterol, two substances required by the human body for proper functioning. Caffeine has been shown to irritate the colon which can lead to problems in the lower digestive tract. Low levels of essential fatty acids can lead to skin disorders, and a low level of gamma linoleic acid (GLA), an omega-6 fatty acid, can manifest itself as inflammation of the gums. Problems in the upper part of the gastrointestinal tract can actually lead to problems in the colon. Improperly digested food that makes it to the colon may cause flatulence because of the bacterial action on the food, and if the normal flora in the colon is out of balance, irritation of the ascending colon can result.[7]

Food allergies have been noted to possibly contribute to IBD. Foods that have been found to be problems are milk, wheat, peanuts, corn and

---

[6] Lipski, 32.
[7] Gary Null, <u>Healing Your Body Naturally: Alternative Treatments to Illness</u> (New York: Seven Stories Press, 1997) 321.

carrageen-containing foods. A condition called leaky gut syndrome also plays a role in the development of food allergies and is discussed later. Irritation of the descending colon can possibly indicate a food allergy or intolerance[8], and symptoms include nausea, vomiting, diarrhea and abdominal pain.

Stress has been noted to adversely affect digestive function and is known to aggravate both IBD and IBS. Excess acid production occurs when there is excessive psychological stress, and this can lead to poor digestion.

Bacterial or viral infections can cause gastroenteritis which is an inflammation of the digestive tract resulting in vomiting and/or diarrhea. In addition, these microorganisms can release toxins in the GI tract that can lead to leaky gut syndrome, a condition seen in both IBD and IBS. Dysbiosis, or an imbalance of the normal flora in the GI tract, has contributed to many different digestive disorders.

Too high or too low hydrochloric acid (HCl) levels also play a role in digestive problems. HCl levels that are too high can lead to heartburn, and levels that are too low can result in inefficient digestion and poor absorption of vitamins and minerals.

Exercise is very important in promoting efficient digestion as it promotes enzyme and HCl production. It also helps to prevent constipation by helping to regulate peristalsis.

Although the focus of this book is the role of fatty acids in the treatment of these digestive disorders, it can clearly be seen that by only altering one factor in the treatment of these disorders is futile. It is in the best interest of the patient to understand that certain foods and lifestyle habits may be contributing to the problem and to look closely at his/her diet and modify it accordingly.

---

[8] Null, 321.

Irritable Bowel Syndrome

Irritable bowel syndrome (IBS) is a functional disorder of the large intestine. There is no evidence of any physical damage to the intestinal wall. Fifteen to twenty percent of the general population has symptoms of IBS with two times as many women affected as men.[9] This disorder usually occurs between the ages of twenty and forty.[10] Eight to seventeen percent of the population has actually been diagnosed with IBS, and only the common cold accounts for more sick days.[11] Half of the episodes have been linked to stress. There does not appear to be an immune factor in IBS; therefore, food sensitivities rather than food allergies may play a role. Traditional physicians usually look for evidence of tumors or polyps, etc. in the intestinal tract but are unlikely to treat under activity or suboptimal functioning of the gastrointestinal tract. This is why IBS has been so difficult to diagnose. The following are possible symptoms of IBS:

> Abdominal pain, usually on left side of abdomen, can be persistent, dull or severe (doubling over), relief sometimes occurring after bowel movement
>
> Gas
> Bloating
> Constipation ("rabbit pellet" stool) and/or diarrhea
> Nausea
> Anxiety
> Depression
> Passing mucous with or without stool
> Heartburn
> Loss of appetite
> Belching
> Rumbling/gurgling in the intestines
> Frequent urination
> Feeling of incomplete evacuation of the bowel
> Fatigue
> Back pain
> Sleeping difficulty
> Headache
> Painful menstruation

---

[9] Rosemary Nicol, <u>Irritable Bowel Syndrome: A Natural Approach</u> (Berkeley: Ulysses Press, 1999) 1-2.
[10] Nicol, 1.
[11] Nicol, 2.

In addition to stress and food sensitivities, another possible factor which is thought to play a role in this disorder is excessive intake of fats. According to Dr. James Braly, foods that should be avoided include nuts, seeds, fruits with seeds, caffeine, alcohol and spices.[12] Supplements that help include vitamin A, zinc and evening primrose oil. Useful herbs include enteric coated peppermint, ginger, chamomile, rosemary and balm. Stress reduction through the use of biofeedback also appears to help.

As will be discussed later, recent research has indicated that a malfunctioning of the serotonin system in the GI tract may play a significant role in this disorder. Some cases of IBS have responded to selective serotonin re-uptake inhibitor (SSRI) antidepressants. Since ninety five percent of the serotonin is located in the GI tract, this finding could be very useful in the treatment of IBS.

Although not life threatening, IBS is a distressing problem and can seriously disrupt a person's life. Because it is a functional disorder and no obvious signs of inflammation or ulceration are found during testing, it has been very difficult to diagnose. This difficulty in the diagnosis can be very frustrating to the patient as he/she wants to know what is causing the discomfort. It is so important to find the cause of this disorder and to find an effective form of treatment. Omega-3 fatty acids may play an important role here since they have been shown to help regulate the serotonin pathway, and this will be discussed in further detail later.

---

[12] Goldberg, 685.

Inflammatory Bowel Disease

Inflammatory bowel disease (IBD) includes both Crohn's disease and ulcerative colitis. It is most common in developed countries and is more likely to occur in higher socioeconomic areas. There are three times as many incidents of this disease in Ashkenazi Jews than non-Jews, and it is much less common in African Americans. Orientals are unlikely to have IBD if they live in the Far East; however, they are more likely to develop the disorder if they move to North America. Both diseases involve inflammation and ulceration of the digestive tract.

Crohn's disease can affect any part of the gastrointestinal tract. Other names for this disease include ileitis, terminal ileitis, regional enteritis, regional ileitis, granulomatous ileitis, granulomatous enteritis, and granulomatous colitis. Forty five percent of cases involve the ileum and the beginning of the colon. Thirty five percent occur in the end of the ileum, and twenty percent occur in just the colon. Twenty five percent have an area of swelling in the lower right abdomen, and twenty five percent have perianal disease in which there are swollen skin tags around the anus along with fistula formation. However, the numbers change somewhat in children. In fifty to seventy percent of children, the disease involves the terminal ileum. The entire bowel wall, from the mucosa to the serosa, is inflamed along with the mesentry. The inflammation may be patchy or continuous, and there is no known cure. Possible symptoms include the following:

>   Abdominal pain with cramping
>   Diarrhea and weight loss
>   Pain in the lower right abdomen
>   Pain occurring after a meal
>   Fatigue
>   Poor appetite
>   Joint pain
>   Rectal bleeding
>   Persistent symptoms
>   Vitamin deficiencies
>   Anemia

Crohn's disease occurs frequently in childhood and adolescents, affecting twenty five to thirty percent of children under the age of twenty.[13]

---

[13] Petar Mamula M.D. and Jonathan Markowitz M.D., "Crohn Disease" eMedicine Journal Volume 3 Number 1 (16 Jan 2002). 15 Jun 2002 <http://www.emedicine.com/ped/topic507.htm>

There is some suggestion that genetics may play a role. Mutations of a gene called NOD2 are thought to play a role in twenty percent of cases.[14]

The risk of Crohn's disease is increased in smokers. Smoking is associated with a higher rate of relapse, repeated surgery and the use of immunosuppressive drugs such as corticosteroids. Interestingly, cigarette smoking has been associated with a low risk of developing ulcerative colitis.

Ulcerative colitis, also known as idiopathic ulcerative colitis and nonspecific ulcerative colitis, affects only the colon. The disease always involves the rectum and is continuous without patches of inflammation. Only the mucosa is inflamed, and the wall bleeds easily. Possible symptoms of ulcerative colitis include:

    Bloody diarrhea
    Tenesmus (spasm of the rectum leading to the feeling of constantly
        having to have a bowel movement)
    Crampy pain in the left side of the abdomen
    Decreased appetite
    Fever
    Chills
    Sweats
    Weight loss

Treatments for IBD usually include an elimination diet to see if food allergies exist. Quercetin, a type of plant pigment known as a flavonoid, helps to reduce inflammation, and chamomile and peppermint are helpful for gas and colic. During the active stages of the disease, fiber is too harsh, and its use is contraindicated; however, when the disease is in remission, fiber is helpful for the prevention of reoccurrence. Dr. Virenden Sodhi suggests the use of flaxseed oil, fish oil and the herb boswellia to help reduce inflammation.[15]

High fat and sugar consumption have been seen in patients suffering from IBD. According to a 1997 study by Reif et al., a diet high in animal fat and cholesterol was associated with an increased risk of UC. This study also

---

[14] Mamula.
[15] Goldberg, 685.

showed a negative association between high intake of fluids, magnesium, vitamin C, fruits and the development of IBD.[16]

Researchers in Tokyo also noted that animal fat plays a role in the development of Crohn's disease. Their study which was published in the American Journal of Clinical Nutrition showed that "…increased incidence of Crohn's disease was strongly correlated with increased dietary intake of total fat, especially saturated animal fat, and omega-6 fatty acids…animal protein, and milk protein; and a low intake of omega-3 fatty acids."[17]

Both ulcerative colitis and Crohn's disease are two serious disorders of the digestive tract and require a physician's care, preferably a gastroenterologist. Since these disorders involve inflammation of the digestive tract, nutrients from food are not properly absorbed. This lack of nutrient absorption affects the entire body and can cause any number of health problems, even those appearing to be separate from the GI tract. Diet appears to play an important role, and saturated animal fat and sugar appear to be major contributors to the development of these digestive ailments. It is imperative to reduce inflammation and to focus on the health of the intestinal wall in the treatment of these diseases. As will be shown later, omega-3 fatty acids may be an effective way to reduce inflammation by increasing anti-inflammatory substances in the body.

---

[16] S. Reif et al, "Pre-illness Dietary Factors in Inflammatory Bowel Disease" Gut Vol 40 (1997). 26 May 2002 <http://gut.bmjjournals.com/cgi/content/abstract>

[17] Flax Oil May Aid Persons with Crohn's Disease 2003. Barleans Organic Oils L.L.C. 16 Dec 2003 <http://www.barleans.com/literature/flax/111-flax-and-crohns.html>

Leaky Gut Syndrome

It is very common to have leaky gut syndrome, or increased intestinal permeability, in cases of IBS and IBD. This syndrome is caused by taking too many pain relievers (NSAIDs), high stress, a junk food diet, environmental toxins and excessive intake of coffee and alcohol. The digestive epithelium repairs and replaces itself every three to five days. Any of the above factors can block the ability of the digestive system to repair itself. The nutrients in foods are absorbed through the intestinal wall and transported to the bloodstream by carrier molecules through a process called active transport. In between these cells are desmosomes with tight junctions which prevent larger substances from passing through. In leaky gut, these tight junctions loosen up. Cells get weak and "holes" appear in the digestive lining. As a result, food remnants get absorbed into the bloodstream. These remnants cannot be utilized by the cells and may actually cause an allergic reaction. These larger particles activate antibodies and cytokines which in turn activate lymphocytes leading to inflammation.

One way to test for leaky gut is to ingest a drink with two sugars in it, one with a small molecular size and the other with a large molecular size. Urine samples are taken over the next six hours. In a person with a healthy digestive tract, high levels of the smaller molecular sized sugar and very small amounts of the larger sized sugar will be seen. However, in patients with digestive problems, high levels of the larger sized sugar may be seen which is indicative of leaky gut syndrome.

As you can see, leaky gut syndrome appears to play a significant role in both IBS and IBD. Since a junk food diet along with coffee and alcohol has been reported to contribute to this problem, diet is once again an important factor in the progress of both of these diseases. A junk food diet is likely to be high in omega-6 fatty acids which produces inflammation and swelling. Therefore, the omega-3 fatty acids may help to counteract this effect through their anti-inflammatory properties.

Essential Fatty Acids

"So profoundly deficient is the American diet in omega-3 fatty acids that a whole host of modern illnesses can be averted, delayed and substantially cured by the addition of these good fats to the diet."[18]

Essential fatty acids (EFA's) are those that cannot be made by the body and must be included in the diet. Linoleic acid (LA) and alpha linolenic acid (LNA) are two of the essential fatty acids. The placement of the double bond in the carbon chain of the fatty acid determines whether or not a fatty acid belongs to the omega-3, omega-6, or omega-9 family. The tail end of the fatty acid is referred to as the omega end, and the type of fatty acid is determined by the location of the double bond in relation to the omega end. Linoleic acid is an 18 carbon chain with two double bonds, the first of which occurs at the sixth carbon; therefore, it is a member of the omega-6 family as illustrated below:

```
                                            O
                                            ||
   C   C   C=C  C=C   C   C   C   C--OH
  / \ / \ /    \ /   \ / \ / \ / \ /
   C   C   C    C     C   C   C   C

Omega      1st double bond        Alpha
end        occurs at 6th carbon   end
```

Linoleic acid (LA)

---

[18] A.P. Simopoulos, "Omega-3 Fatty Acids in Inflammation and Autoimmune Diseases" <u>Journal of the American College of Nutrition</u> 21(6) (Dec 2002). 16 Dec 2003
<http://www.ncbi.nlm.nih.gov/entrez/query.fcgi?cmd>

Other examples of omega-6 fatty acids include gamma linoleic acid (GLA), arachidonic acid (AA) and dihomo-linoleic acid. AA and GLA are illustrated below:

```
                    O
                    ||
      C=C   C=C   C   C--OH           C=C   C   C
     / \ / \ / \ /                   / \ / \ / \
    /   C    C   C                  /   C   C   C--OH
   C                               /            ||
    \                             C             O
     \   C    C   C   C  Omega    \
      \ / \ / \ / \ /    end       \   C    C   C   C  Omega
       C=C   C=C   C   C            \ / \ / \ / \ /    end
                                     C=C   C=C   C   C
      Arachidonic Acid
           (AA)                        Gamma linoleic acid
                                              (GLA)
```

On the other hand, alpha linolenic acid is an 18 carbon chain with 3 double bonds with the first occurring at the third carbon. This fatty acid belongs to the omega-3 family:

```
                                                    O
                                                    ||
    C   C=C   C=C   C=C   C   C   C   C--OH
     \ / \ / \ / \ / \ / \ / \ / \ /
      C    C    C    C   C   C   C

   Omega       1st double bond              Alpha
   end         occurs at the 3rd carbon     end
```

Alpha linolenic acid
(LNA)

Other examples of omega-3 fatty acids are eicosapentenoic acid (EPA) and docosahexanoic acid (DHA) which are illustrated below. Notice the location of the double bond in relation to the omega end of the chain.

```
                O
                ||
    C=C   C=C   C   C--OH            C=C   C=C   C=C   C
   / \  / \  /\ /                   / \  / \  / \  /\
  /    C    C   C                  /    C    C    C   C--OH
 C                                /                    ||
  \    C    C    C               C                     O
   \  / \  / \   / \              \
    C=C   C=C   C=C  C             \    C    C    C
                                    \  / \  / \  / \
     Eicosapentenoic acid            C=C   C=C   C=C   C
            (EPA)
                                        Docosahexanoic acid
                                               (DHA)
```

An omega-9 fatty acid, called oleic acid, is monounsaturated and found in olive oil, avocados, and nuts. Note below that the double bond occurs at the 9th carbon.

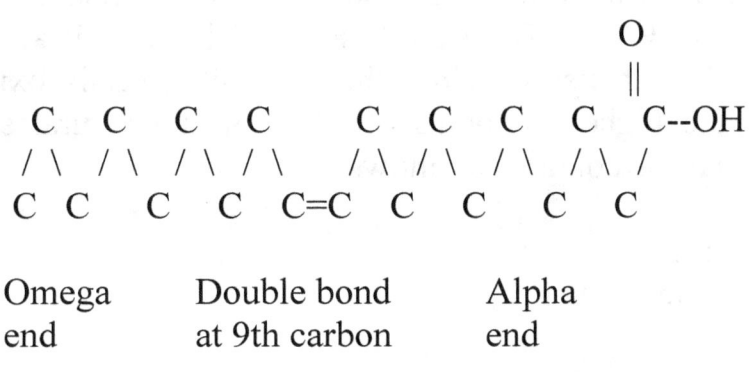

Omega   Double bond   Alpha
end   at 9th carbon   end

Oleic acid

The omega-3 fatty acids are extremely important in the body. In addition to being precursors to prostaglandins, they help to keep cell membranes fluid and flexible which allows for effective exchange of nutrients in and out of the cell. Alpha linolenic acid, found primarily in plant foods, is converted in the body to EPA and DHA. Both EPA and DHA are critical for building neural tissue, but it appears that only DHA can stimulate the actual growth of nerve cells. DHA can actually cause the number of neuronal connections to increase.

Several vitamins and minerals are necessary for the omega-3's to do their job and to help convert these fatty acids to prostaglandins. These include vitamins A, B, C, E and the minerals calcium, selenium, copper, zinc, and manganese. The B vitamins help to boost the action of omega-3 fatty acids, and vitamin B6 is pivotal in helping to make prostaglandins; however, taking large doses of vitamin B with omega 3's should be avoided as this may greatly increase the effectiveness of this vitamin. Vitamin E helps to protect the EFA's from oxidation; calcium helps to stabilize nerve conduction; and selenium guards the fats in cell membranes.

Omega-3's should never be used to deep fry foods because the oil is destroyed resulting in the production of toxic substances. Because these fatty acids are unsaturated, they will oxidize and turn rancid easily. As a result, the modern food industry has come up with a process called hydrogenation which saturates these fats, making them more stable with a longer shelf life. This process has destroyed the effectiveness of these fatty acids and has led to a drastic reduction in these vital fats in the modern day diet.

Omega-3 fatty acids can be found in both flaxseed and fish oil. LNA is found in flaxseed, and EPA and DHA are found in fish oil. It is advisable to take extra Vitamin E with fish oil since this product is easily oxidized. Flaxseed has one of the highest amounts of LNA in any natural source. Other components of flaxseed oil are as follows:

16% linoleic acid (LA)
18% oleic acid (omega-9)
Vitamin E
Beta carotene

Other components of flaxseed meal are:

    Calcium
    Potassium
    Magnesium
    Manganese
    Soluble fiber
    Lignins (cancer fighter)
    Protein

Other good food sources of omega-3's are sardines, salmon, mackerel, herring, striped bass, soybeans, walnuts, and dark leafy green vegetables. A small plant with succulent leaves called purslane, often considered a pest for gardeners, is also an excellent source of omega-3 fatty acids.

There is some disagreement over whether flaxseed (linseed) oil or fish oil is better in obtaining high levels of omega-3 fatty acids. LNA inhibits the delta 6-desaturase enzyme, and therefore the conversion of LNA to EPA and DHA is limited. However, a study done in Germany determined that "a diet of linseed oil (30 ml. daily) for four weeks raised the content of LNA by 200 percent, the level of EPA by 150 percent and the level of DHA by 70 percent. The human body seems capable of transforming linolenic acid into EPA. This change coincided with a significantly reduced production of platelet thromboxanes (clotting agents)."[19] Also, according to Dr. Andrew Weil, "if the diet is top-heavy in omega-6 fatty acids, those will compete for a necessary enzyme, blocking the synthesis of DHA."[20] In addition, there is also some concern about possible toxins in fish oil. These concerns need to be studied further to determine just how much of an effect these factors will have on each supplement.

The omega-6 fatty acids are important in the human body since they are also precursors to prostaglandins. They are the most common polyunsaturated fat found in foods. Gamma linoleic acid (GLA) is the most useful of these fatty acids in treating health problems since this omega-6 fatty acid can lead to the production of "good" eicosanoids. However, too much GLA can also lead to the production of "bad" eicosanoids, such as AA, leading to pain, swelling and inflammation. This will become clearer in

---

[19] Ingeborg M. Johnston, C.N. and James R. Johnston, Ph.D. Flaxseed (Linseed) Oil and the Power of Omega-3 (Los Angeles: Keats Publishing, 1990) 22.
[20] Andrew Weil, M.D., Eating Well for Optimum Health (New York: Alfred A. Knopf, 2000) 88.

the chapter on eicosanoids where the actual pathways will be discussed. LA can be found in high amounts in seeds, corn and fish in warmer waters.

The National Institute of Health recommends the following daily intake of essential fatty acids:[21]

EPA/DHA 650mg.
LNA 2.22 g.
LA 4.44 g.

In addition to these long chain fatty acids, there are short chain fatty acids that also play a role in the digestive tract. Although these will just be mentioned here, their role in the proper functioning of the GI tract is vital. Some examples include butyric acid, propionic acid, acetic acid and valerate. These fatty acids are produced through the fermentation of fiber by intestinal bacteria and are the main source of fuel for the cells in the colon. Butyric acid specifically has been shown to heal inflammation in bowel tissue, and low levels of this short chain fatty acid have been seen in IBD. The chemical structure of butyric acid is shown below:

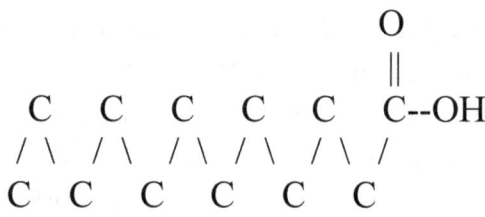

Butyric acid

Both omega-3 and omega-6 fatty acids are necessary in the human body. The omega-3 fatty acids include LNA, EPA and DHA, and the omega-6's include LA, GLA and AA. Both LA and LNA are building blocks of other fats. Since both are precursors to prostaglandins, they can increase the effectiveness of the immune system and help to control inflammation. Omega-6 fatty acids can cause swelling, inflammation and constriction of blood vessels, and this effect is needed at times to help

---

[21] Angela Rank, Eicosanoids Wake Forest University, 10 Jan 2004
<http://www.wfu.edu/users/clafme0/nutrition/eicosanoids.htm>

prevent other health problems. An example is when a person is cut and is bleeding. In this instance, it is necessary for the blood vessel to constrict and form a clot to help stop the bleeding. If there was a low level of omega-6 fatty aids in the body (along with other nutrient deficiencies) and the blood vessels did not constrict, the person could bleed to death. However, if there are too many omega-6 fatty acids present in the body and not enough omega-3's, there would be too much swelling, inflammation, blood vessel constriction, etc. which may lead to other health problems such as cardiovascular disease or inflammatory conditions such as Crohn's disease and ulcerative colitis. Omega-3 fatty acids promote vasodilation of blood vessels and have anti-inflammatory properties. It is vitally important to have a balance of omega-3 and omega-6 fatty acids in the diet, and since the modern day diet is so low in omega-3's, there needs to be a genuine effort to increase our intake of these essential fatty acids.

Eicosanoids

Eicosanoids are a type of hormone that requires polyunsaturated fat for their synthesis and have over one hundred known functions. These substances have both helpful and harmful effects on the body, and it is important that they are balanced in the body. "Good" eicosanoids prevent blood clots, dilate blood vessels, reduce pain, enhance the immune system and improve brain function. "Bad" eicosanoids promote blood clots, constrict blood vessels, promote pain, and decrease immune and brain functions. For example, some eicosanoids stimulate blood clotting while others decrease it. The stimulation of blood clotting is important when someone is bleeding as this will help stop the bleeding; however, if the blood clots too much, this can lead to possible cardiovascular disease or other health issues. The eicosanoid which decreases blood clotting counteracts the one that stimulates the clotting, so again, balance is important. Other known functions of eicosanoids include regulation of blood pressure, control of allergic responses and regulation of smooth muscle contraction. The different groups of eicosanoids include the prostaglandins (PG) which regulate pain and inflammation, prostacyclins (PGI) which decrease blood clotting, thromboxanes (TX) which stimulate blood clotting, leukotrienes (LT) which control the inflammatory response and lipoxins (LX) which play a role in immune and allergic responses. An increase in "bad" eicosanoids causes inflammation through proteins produced by immune cells called cytokines. These cytokines can lead to the production of more "bad" eicosanoids and a vicious cycle starts. "Eicosanoid control has 90% of the impact on the pain you feel."[22]

As previously mentioned, both omega-3's and omega-6's are precursors to prostaglandins. Prostaglandins are a type of eicosanoid which regulate pain, inflammation and swelling. They play a role in digestion, blood pressure control, heart function, kidney function, allergic reactions, blood clotting and hormone production. Omega-3's make certain kinds of prostaglandins while the omega-6's make others. LNA and EPA produce the "good" eicosanoids, also referred to as series I eicosanoids, while the LA produces the "bad" eicosanoids, or series II eicosanoids because it is the precursor to AA. However, "good" eicosanoids can also be produced through the LA pathway through the action of dihomogamma linolenic acid (DGLA). DGLA leads to the production of series I eicosanoids. The process by which LA metabolizes into DGLA and AA is shown later. EPA is converted into the series III eicosanoids which have been shown to lower

---

[22] Sears, 169.

triglycerides, reduce pain and inflammation from rheumatoid arthritis and inhibit the production of AA. Although the omega-6's produce the "bad" kind, they are necessary for the proper functioning of the body. The levels of prostaglandins in the body can be changed by changing the intake of essential fatty acids. There must be a balance of both good and bad eicosanoids for proper health.

The regulation of eicosanoids is influenced by the following factors:

-High insulin shifts the balance toward the "bad" eicosanoids.
-Trans-fatty acids shift the balance toward the "bad" eicosanoids.
-Glucagon, a hormone released from the pancreas that increases blood sugar levels, shifts the balance toward the "good" eicosanoids.
-LNA (flaxseed), EPA and DHA (fish) shift the balance toward the "good" eicosanoids.

Once fatty acids are released from cell membranes through the action of phospholipase A2, they follow one of three pathways: cyclo-oxygenase, 5-lipoxygenase or 12 or 15-lipoxygenase. The cyclo-oxygenase pathway, or COX, produces thromboxanes and prostaglandins. There are two types of COX enzymes, COX-1 and COX-2. COX-1 enzymes are in vascular cells that line the bloodstream and stomach and secrete bicarbonate which neutralizes stomach acid. The COX-2 enzyme is released in response to inflammation. The 5-lipoxygenase, or 5-LIPO system leads to production of leukotrienes. The 12 or 15-lipoxygenase pathway leads to production of hydroxylated fatty acids and lipoxins. There are very few inhibitors of the LIPO enzymes. Corticosteroids work but have many side effects. In addition, cortisol, released from the adrenal gland as a result of stress, synthesizes lipocortin which inhibits the action of phospholipase A2 in an attempt to stop production of "bad" eicosanoids. So excess cortisol, in preventing the release of fatty acids from cell membranes, can depress the immune system, disrupt short term memory and destroy nerve cells in the brain.

Fatty acids must be at least 20 carbons in length with at least three double bonds in order to be converted into an eicosanoid. Linoleic acid begins as an 18 carbon length chain but is converted into a 20 carbon chain (DGLA) through the following process:

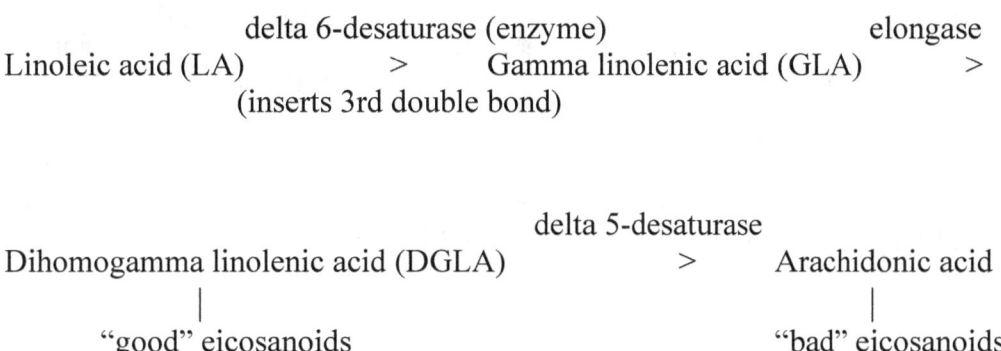

The balance of DGLA and AA will determine what kind of eicosanoid will be produced. An eicosanoid produced from DGLA is the prostaglandin PGE1 which is a good vasodilator, inhibitor of platelet aggregation and a reducer of insulin secretion. Another powerful eicosanoid produced from DGLA is PGA1. This is a strong suppressor of viral replication, and it inhibits NF kappa B which reduces inflammation by suppressing the production of pro-inflammatory cytokines. DGLA also leads to the production of 15-HETriE which is a potent inhibitor of leukotriene synthesis. High levels of insulin activate delta 5-desaturase leading to production of more AA and "bad" eicosanoids. AA promotes an increased release of pro-inflammatory cytokines. This fatty acid produces the prostaglandin PGE2 and the leukotriene LTB4, both of which promote inflammation. Other "bad" eicosanoids produced through the AA pathway include the lipoxins A4 and B4 and the prostacyclin PG12. In addition, AA produces TXA2 which decreases blood flow.

The process by which LNA is converted to eicosanoids is summarized below:

                                  delta 6-desaturase   elongase (enzyme)
Alpha linolenic acid (LNA)   >   Steradonic acid   >

                                  delta 5-desaturase   elongase x 2
Eicosatretaeonic acid   >   eicosapentenoic acid (EPA)   >

                                  delta 6-desaturase   peroxisomal oxidation
24 long carbon chain,   >   24 long carbon chain,   >
5 double bonds   6 double bonds

                        peroxisomal oxidation
docosahexanoic acid (DHA)   >   eicosapentenoic acid (EPA) - see above
                                                Process is repeated to produce more DHA

EPA inhibits the production of AA thereby reducing the production of "bad" eicosanoids. EPA also inhibits the delta 5-desaturase enzyme, promoting a more balanced ratio of DGLA to AA in the LA pathway.

Because of the high intake of omega-6 fatty acids in the modern day diet, the production of AA leading to the release of pro-inflammatory cytokines is so high that it may be leading to inflammatory conditions such as Crohn's disease and ulcerative colitis. As noted above, both types of eicosanoids are needed in the right balance for optimum health. The modern day diet, with all of the processed foods, promotes the overproduction of the "bad" eicosanoids while the amount of omega-3 fatty acids, which lead to the production of "good" eicosanoids, has dropped dramatically. For this reason, it is so important to increase our intake of omega-3 fatty acids and make a genuine effort to reduce our intake of hydrogenated products and trans fats.

The Role of Cytokines in IBD

The overproduction of pro-inflammatory cytokines in the body is a direct result of consuming too many omega-6 fatty acids and not enough omega-3's. This leads to a shift toward the production of the "bad" eicosanoids which cause the production of the pro-inflammatory cytokines. Tissue in the digestive tract called GALT or gut associated lymphatic tissue plays a role in the production of these cytokines. Foreign substances called antigens are carried by M-cells to Peyer's patches located in the intestinal lining. The cells in these patches alert the B- and T-cells of the immune system which respond and carry the antigen to the mucosa where they are "eaten" up by macrophages. When this occurs, secretory IgA is aroused in the mucosa and begins the inflammatory process by activating cytokines.

Crohn's disease and ulcerative colitis are both characterized by high levels of the pro-inflammatory cytokines called interleukin 1 (IL-1), IL-6, and leukotriene LTB4. In intestinal lesions, IL-2, interferon (IFN) and tumor necrosis factor (TNF) are seen in Crohn's disease while IL-5 is predominant in ulcerative colitis. In a study by Wang et al., mice in which ulcerative colitis had been induced were tested for levels of pro-inflammatory cytokines. They found that there was an increase in TNF-alpha and IL-1 along with an increase in the activity of NF-kappa B after the onset of colitis. TNF-alpha is a protein that produces inflammation as part of the normal immune response. Overproduction of this protein can lead to excessive inflammation. NF-kappa B is a transcription factor that activates certain genes, and this activation can lead to inflammation and the destruction of tissue.[23] They concluded "pro-inflammatory cytokines play important roles in the pathogenesis of UC and may exacerbate the inflammation of the intestinal mucosa and cause apoptosis of epithelial cells, possibly under the regulation of NF-kappa B activation."[24] Omega-3 fatty acids can help shift the production of eicosanoids to the "good" side resulting in fewer pro-inflammatory cytokines leading to less pain and inflammation.

Anti-inflammatory cytokines have been shown to have a beneficial effect on patients with IBD. In a study by Torsten et al, the effects of anti-inflammatory cytokines were tested to determine their effects on patients

---

[23] V. Gilston, "Inflammatory Mediators, Free Radicals and Gene Transcription" Progress in Inflammation Research (1999) 28 Feb 2004 <http://www.birkhauser.ch/books/biosc/pir/pir5851toc.html>

[24] QY Wang et al, "Expression of Pro-Inflammation Cytokines and Activation of Nuclear Factor Kappa B in the Intestinal Mucosa of Mice with Ulcerative Colitis" Di Yi Jun Yi Da Xue Xue Bao 23(11) (Nov 2003). 14 Jan 2004 <http://www.ncbi.nlm.nih.gov/entrez/query.fcgi?cmd>

with IBD. An increase in pro-inflammatory cytokines is seen in IBD, and anti-inflammatory cytokines, such as IL-4, IL-10, and IL-13 were tested to see if these can help in reducing inflammation. The combination of IL-10/IL-4 and the combination of IL-10/IL-13 showed an inhibition of the pro-inflammatory cytokines. They summarized that "a combination of anti-inflammatory cytokines is more effective in down-regulating the response of activated monocytes than using the cytokines alone and this may have a potential therapeutic benefit for patients with IBD."[25]

A genetic component was suggested as a factor in IBD in another study by Tagore et al. This genetic link was found to be linked to a low level of the anti-inflammatory cytokine IL-10. Genotyping was performed on blood cells from IBD patients, and there was found to be a low level of the high IL-10 producer allele present in these patients. This study "…suggests that individuals genetically predisposed to produce less IL-10 are at a higher risk of developing IBD, in particular, UC."[26]

There is a clear link between the pro-inflammatory cytokines and IBD. The presence of too many of these cytokines is directly related to the amounts of omega-6 fatty acids that are in the diet since omega-6's lead to the production of AA. Anti-inflammatory cytokines have been shown to reduce inflammation and have been found to be beneficial in the treatment of IBD. Since omega-3 fatty acids reduce the number of pro-inflammatory cytokines, these EFA's should be included in any treatment plan for these digestive disorders. The anti-inflammatory benefits of these EFA's are discussed in the next chapter.

---

[25] Torsten Kucharzik, "Synergistic Effect of Immunoregulatory Cytokines on Peripheral Blood Monocytes from Patients with Inflammatory Bowel Disease" <u>Digestive Diseases and Sciences</u> 42(4) (1 Apr 1997). 14 Jan 2004 <http://www.ncbi.nlm.nih.gov/entrez/query.fcgi?cmd>

[26] A. Tagore et al, "Interleukin-10 (IL-10) Genotypes in Inflammatory Bowel Disease" <u>Tissue Antigens</u> Volume 54, Issue 4 (Oct 1999) 14 Jan 2004 <http://www.blackwell-synergy.com/servlet/useragent?func>

Anti-inflammatory Effects of Omega-3 Fatty Acids in IBD

"Recent research shows that over 60 percent of the immune system is located in or around the digestive system."[27]

Omega-3 fatty acids have been shown to have a dramatic impact on the functioning of the immune system. Specifically EPA and DHA have been shown to have more of an effect that LNA. The omega-3 fatty acids found in fish and flaxseed oil are known to reduce the production of leukotriene B4 and thromboxane A2. They are also known to inhibit the synthesis of pro-inflammatory cytokines. The anti-inflammatory effects of these substances have been found to be useful in patients with IBD. The actions of these fatty acids are noted below:

-Omega-3's determine the types of eicosanoids that are produced.
-They have been shown to act upon intracellular signaling pathways and gene expression.

Crohn's disease and ulcerative colitis have an increase in the levels of interleukin 1 (IL-1) and leukotriene B4 (LTB4). Both of these substances are produced from the omega-6 family of fatty acids. Many different studies have shown that the use of omega-3 fatty acids in the treatment of these diseases resulted in decreased disease activity and reduction of use of anti-inflammatory drugs.

In a study by A. Asian and G. Triadafilopoulos, published in the American Journal of Gastroenterology in 1992, eleven patients with mild to moderate ulcerative colitis were studied in a double-blind placebo controlled trial in which their diet was supplemented with fish oil delivering 4.2 g. of omega-3 fatty acids per day. Symptoms were assessed by patient reports and sigmoidoscopy, and leukotriene B4 levels were monitored by immunoassay. There was a 56% reduction in disease activity for those receiving fish oil as opposed to only a 4% reduction in those receiving the placebo. However, there were no significant differences in leukotriene B4 levels. Eight patients receiving the fish oil were able to reduce or eliminate anti-inflammatory drugs.[28]

---

[27] Lipski,7.
[28] A. Asian and G. Triadafilopoulos, "Fish Oil Fatty Acid Supplementation in Active Ulcerative Colitis: A Double-Blind, Placebo-Controlled, Crossover Study" American Journal of Gastroenterology 87/4 (1992). 21 May 2002 <http://www.lef.org/prod_hp/abstracts/php-ab161b.html>

Although leukotriene B4 levels were not significantly altered in the previous study, Hillier et al. showed that other eicosanoids could be altered by the use of fish oil, specifically prostaglandins and thromboxanes. During a 12 week study, one group received 18 g. of fish oil daily while the other group received olive oil. In the fish oil group, EPA rose seven fold and DHA rose 1.5 fold after three weeks of supplementation. At 12 weeks, it was noted that the AA levels had fallen significantly. Reductions in PGE2 were also seen at 3 and 12 weeks, and thromboxane B2 levels fell significantly at 12 weeks. Since the main fatty acid in olive oil is oleic acid (omega-9), it is not surprising that a significant increase in oleic acid was seen at 12 weeks in the group receiving olive oil. There was also a fall in DHA in the group who took olive oil along with an insignificant change in eicosanoid synthesis.[29]

In one of the better known studies involving the use of fish oil in the treatment of Crohn's disease, a dramatic increase in the rate of remission was seen. This study, performed by Belluzzi et al. in Italy, monitored two groups of patients that had been in remission for one year. One group received nine fish oil capsules daily, each containing 2.7 g. of omega-3 fatty acids, while the other group received nine placebo pills daily. After one year, 59 percent of the fish oil group was still in remission as compared to only 26 percent of the placebo group. The researchers concluded "In patients with Crohn's disease in remission, a novel enteric-coated fish-oil preparation is effective in reducing the rate of relapse."[30]

In a 6 month study published in the American Journal of Gastroenterology in 1988, dramatic results were seen as a result of fish oil supplementation in patients with ulcerative colitis. Researchers divided ulcerative colitis patients into two groups. The first group received 15 ml. of fish oil per day while the other group received 15 ml. of sunflower oil. The fish oil group had a significant decrease in their symptoms, and after 3 months, all of them went into remission. The placebo group did not improve, and some patients worsened, requiring treatment with steroids. It was determined through clinical examination that the EFA's suppressed the cytotoxic activity of the killer cells in this disease.[31]

---

[29] K. Hillier et al, "Incorporation of Fatty Acids from Fish Oil and Olive Oil into Colonic Mucosal Lipid and Effects upon Eicosanoid Synthesis in Inflammatory Bowel Disease" Gut 32/10 (1991). 21 May 2002 <http://www.lef.org/prod_hp/abstracts/php-ab161b.html>

[30] Andrea Belluzzi et al., "Effect of an Enteric-Coated Fish-Oil Preparation on Relapses in Crohn's Disease" The New England Journal of Medicine Volume 334, Number 24 (13 Jun 1996). 23 Feb 2004 <http://content.nejm.org/cgi/content/abstract/334/24/1557?where>

Grimminger et al. worked with a 36 year old female with ulcerative colitis who was suffering from severe side effects from the administration of steroids. They gave her an intravenous form of fish oil containing 4.2 g. EPA and 4.2 g. DHA for 9 days while rapidly reducing her steroid medication. The disease activity declined to the point where she could take the fish oil by mouth. The ulcerative colitis returned during this time, and they went back to the intravenous administration once again. During the intravenous supplementation, marked improvement occurred and EPA and DHA levels increased, surpassing the levels of AA. The researchers concluded that "The profound changes in fatty acid profiles and lipid mediator generation may be related to the reduction in colitis activity observed during the periods of intravenous n-3 lipid supplementation."[32] In this study, it is interesting that the ulcerative colitis returned during the time she was taking the fish oil by mouth. It is worthwhile to note here that her GI tract may still have been very inflamed and absorption of the fish oil would not have been efficient enough at that time to help reduce symptoms. Note that her symptoms subsided during both times of intravenous administration.

A high amount of omega-6 fatty acids with a deficiency of omega-3 appears to play a role in these digestive disorders. An article in Metabolism published in 1996 discussed a study in which two groups were examined to determine the levels of essential fatty acids present. One group had chronic intestinal disorders while the other group served as a control group. The study showed that patients with the intestinal disorders had "…an increased ratio of derivatives to precursors of omega-6 fatty acids, shifts that occur when the cells are EFA-deficient." They concluded that those patients "…with chronic intestinal disease should be evaluated for likely EFA deficiencies and imbalances and treated with substantial amounts of supplements rich in EFA's, such as oral vegetable and fish oils, or intravenous lipids if necessary."[33]

---

[31] <u>IBD and Fatty Acids</u> 2002. Great Smokies Diagnostic Laboratory. 21 May 2002
<http://www.gsdl.com/assessments/finddisease/ibd/fatty_acids.html>

[32] Grimminger et al., "Influence of Intravenous n-3 Lipid Supplementation on Fatty Acid Profiles and Lipid Mediator Generation in a Patient with Severe Ulcerative Colitis" <u>European Journal of Clinical Investigation</u> 23/11 (1993). 21 May 2002 <http://www.lef.org/prod_hp/abstracts/php-ab161b.html>

[33] "Prevalence of Essential Fatty Acid Deficiency in Patients with Chronic Gastrointestinal Disorders" <u>Metabolism</u> 45(1) (Jan 1996). 21 May 2002 <http://www.lef.org/prod_hp/abstracts/php-ab161b.html>

In yet another study conducted at the Clinic of Abdominal and Transplantation Surgery in Hannover Germany, tissue samples of patients with Crohn's disease were tested while using samples from healthy subjects as controls. Patients with the inflamed tissue had low levels of omega-3 fatty acids along with very high levels of AA as compared to the controls.[34]

There is plenty of evidence that omega-3 fatty acids reduce inflammation and reduce disease activity in IBD. This reduction of inflammation appears to be a result of the production of "good" eicosanoids and the reduction of pro-inflammatory cytokines. According to E. Ross at Tufts University School of Medicine, "Fish oils may exert their beneficial effects by shifting eicosanoid synthesis to less inflammatory species or by modulating tissue levels of certain cytokines."[35] Therefore, omega-3 fatty acids should be considered a vital part of the treatment plan in IBD.

---

[34] Flax Oil May Aid Persons with Crohn's Disease
[35] E. Ross, "The Role of Marine Fish Oils in the Treatment of Ulcerative Colitis" Nutrition Review 51/2 (1993). 21 May 2002. <http://www.lef.org/prod_hp/abstracts/php-ab161b.html>

Serotonin and IBS

Serotonin, or 5-hydroxytryptamine (5-HT) is a type of neurotransmitter found in the brain, platelets, and gastrointestinal tract and is a product of the breakdown of tryptophan. It is known to regulate the secretion of digestive fluids and stimulate peristalsis as well as constrict blood vessels. Ninety five percent of the serotonin in the human body is located in the gastrointestinal tract. Many researchers are now looking at serotonin as a possible factor in IBS.

There are three main groups of neurotransmitters: single amino acids, neuropeptides and monoamines. The single amino acids comprise the highest amount of neurotransmitters seen in the human body. The neuropeptides consist of two or more amino acids and are the most powerful. Examples of neuropeptides are the enkephalins and endorphins. The monoamines consist of one amino acid along with various other molecules attached to the amino acid. There are two main families of monoamines: catecholamines such as epinephrine, norepinephrine and dopamine, and the indoleamines such as serotonin and melatonin. Serotonin can act as both a neurotransmitter and a hormone. A neurotransmitter does not travel far and acts on nearby neurons while a hormone is produced in one part of the body and travels to another part to exert its influence. The main neurotransmitters found in the enteric nervous system include GABA, acetylcholine, norepinephrine, serotonin, ATP and nitric oxide (NO).

In order to understand the function of neurotransmitters, it is important to understand how a neuron works. A neuron consists of dendrites, axons and a cell body. An axon can make contact with another neuron at its dendrites, cell body or other axons. A synapse or synaptic cleft is the space where a pre-synaptic axon meets another post-synaptic neuron. As a nerve impulse travels down the pre-synaptic axon, neurotransmitters, such as serotonin, are released into the synapse. Neurons which specifically secrete serotonin are called serotonergic. These molecules then bind with a receptor and the nerve impulse continues to travel through the post synaptic neuron. In addition, there are receptors on the pre-synaptic neuron that act as a "sponge", taking up any neurotransmitter that was not used. This process is called re-uptake. Once a neurotransmitter has done its job, there are three different things that can happen:

1. It can go through re-uptake as just described.
2. It can float around until it binds with another receptor initiating another nerve impulse.

3. It can be cut up by enzymes and be eliminated from the body.

There are seven types of serotonin receptors, some of which have additional subtypes as listed below:

5-HT1 - subtypes 5-HT1A, 5-HT1B, 5-HT1D, 5-HT1E, 5-HT1F
5-HT2 - subtypes 5-HT2A, 5-HT2B, 5-HT2C
5-HT3
5-HT4
5-HT5 - subtypes 5-HT5A, 5-HT5B
5-HT6
5-HT7

5-HT1 and 5-HT2 receptors are primarily located in the brain and are the target for antidepressants. 5-HT3 and 5-HT4 receptors are located in the gastrointestinal tract. 5-HT4 specifically inhibits visceral sensitivity while helping to regulate peristalsis and intestinal secretion. The recalled IBS drug Zelnorm worked by activating this receptor. Antidepressants have some effect on the 5-HT3 and 5-HT4 receptors, and this may explain why antidepressants have helped some patients with IBS.

There are several problems which can occur in the proper functioning of the synapse resulting in the receptor becoming unresponsive to the presence of the neurotransmitter. If the re-uptake system malfunctions, there will not be enough neurotransmitter available for release when the next nerve impulse travels down the axon. If enzyme levels are too high, too much of the neurotransmitter will be destroyed and eliminated from the body. If the receptors are blocked, the neurotransmitter will not be able to bind to its sites and continue the nerve impulse. Overactive neurons may lead to problems by releasing too much neurotransmitter. As can be seen, many different factors can lead to neurotransmitter problems which can cause health problems such as IBS and depression.

In a study by Peter Moses, M.D. and Gary Mawe, PhD. at the University of Vermont, IBS was found to be the result of abnormal alterations in serotonin signaling. The serotonin re-uptake system was found to be diminished in patients with IBS. As a result, the serotonin that is released stays around longer resulting in changes in motility and secretion in the large intestine. They also found higher amounts of endocrine cells, but the amount of serotonin released from these cells was not significantly different than in normal subjects.[36]

Omega-3 fatty acids have been shown to help regulate the levels of serotonin in the brain without the side effects of drugs. Drugs may elevate one neurotransmitter while depressing another neurotransmitter. Fish oil (omega-3 fatty acids) is more effective since it maintains adequate amounts of both neurotransmitters. Two studies follow that demonstrate the effects these fatty acids can have on serotonin levels.

H. Li, D. Liu and E. Zhang divided rats into four groups. The control group received no DHA supplementation while the other three groups received varying amounts of this fatty acid. The serotonin levels in each group were monitored, and the levels of serotonin were found to be significantly higher in the groups receiving DHA than in the control group.[37]

Pakala et al. showed that of all "…fatty acids tested, only EPA and DHA could block the mitogenic effect of serotonin on vascular smooth muscle cells…This antimitogenic effect of EPA and DHA may partially explain some of the beneficial effects of fish oils."[38]

Since 95 percent of serotonin is located in the GI tract and certain serotonin receptors have been shown to help regulate peristalsis, this neurotransmitter may play a vital role in IBS. Positive response of some IBS sufferers to antidepressants further suggests the role of serotonin in IBS. Since this discovery is a relatively new one, more research needs to be done to determine the exact mechanism of the malfunction that may be occurring in this disorder.

---

[36] Jennifer Nachbur, <u>UVM Researchers Identify Molecular Changes in IBS Patients</u> 22 Oct. 2003. The University of Vermont. 10 Jan 2004 <http://www.uvm.edu/news/print/?action=Print&storyID=4188>

[37] H. Li, D. Liu and E. Zhang, "Effect of Fish Oil Supplementation on Fatty Acid Composition and Neurotransmitters of Growing Rats" <u>Wei Sheng Yan Jiu</u> 29(1)(30 Jan 2000). 14 Jan 2004 <http://www.ncbi.nlm.nih.gov/entrez/query.fcgi?cmd>

[38] R. Pakala et al., "Eicosapentaenoic Acid and Docosahexaenoic Acid Block Serotonin Induced Smooth Muscle Cell Proliferation" <u>Arteriosclerosis, Thrombosis and Vascular Biology</u> 19(10) (Oct 1999). 14 Jan 2004 <http://www.ncbi.nlm.nih.gov/entrez/query.fcgi?cmd>

Imbalance of Omega-3 to Omega-6 Fatty Acids - Effects of a Modern Day Diet

"Nearly half of our caloric intake comes from nutritionally depleted foods."[39]

The rate of IBD has been growing steadily in the United States since the 1930's. When cultures switch to a more Westernized diet consisting of highly processed and low fiber foods, the cases of Crohn's and ulcerative colitis increase. "As food-processing techniques removed the vitamins, fiber, and other essential nutrients from our foods, the modern maladies grew rampant."[40]

Rudin states that omega-3 fatty acids have decreased 80 percent in the past 100 years while omega-6 consumption has increased.[41] This imbalance of omega-3 to omega-6 fatty acids leads to unbalanced levels of prostaglandins which can cause health problems. Our prehistoric ancestors were getting about a 1:1 omega-6 to omega-3 ratio in their food; however, that ratio has changed dramatically in recent times. Today's diet has a ratio of 10:1 or even as high as 30:1 omega-6 to omega-3. As you can see, the modern diet supports the production of "bad" eicosanoids because of the large amounts of omega-6 fatty acids that are in our diet. In addition, trans fatty acids can also cause a deficiency of omega-3 fatty acids.

Food processing has added to this problem. In addition to the problems seen with trans fatty acids, essential nutrients are being removed during food processing. For example, according to Elizabeth Lipski, "Whole wheat contains twenty-two vitamins and minerals that are removed to make white flour. After the bran and germ are removed from the whole wheat kernel, so too are 98 percent of pyridoxine (vitamin B6), 91 percent of manganese, 84 percent of magnesium and 87 percent of fiber."[42] Lipski also reports that an average person eats 14 pounds of additives per year.[43] Herbicides and pesticides also affect our health. According to Lipski, "The average person consumes one pound of these chemicals each year. These pesticides have neurotoxic effects and can cause damage to our nervous systems."[44]

---

[39] Lipski, 22.
[40] Donald Rudin, M.D. and Clara Felix, <u>Omega-3 Oils: A Practical Guide</u> (U.S.: Avery, 1996) 2.
[41] Johnston, 19-20.
[42] Lipski, 28.
[43] Lipski, 29.
[44] Lipski, 28.

Poor diet plays a huge role in our health. In a book originally published in 1996, Lipski reports that in the last 10 years, there has been a 10 percent decrease in the amount of vegetables consumed and a 60 percent increase in the number of soft drinks consumed.[45] Some researchers state that 60 percent of calories from today's diet come from refined sugars and fats.[46] It should not be surprising to know that there is such a rise in inflammatory conditions such as Crohn's disease and ulcerative colitis.

Although food processing has helped society in terms of convenience and has resulted in a reduction of food borne illness due to microbial contamination, it has set off a new health dilemma. As previously stated, trans fats produced through food processing have led to a dramatic rise in the amounts of omega-6's consumed while the available omega-3's in food have rapidly declined. It is imperative that we increase our consumption of omega-3 fatty acids and decrease our intake of processed foods.

---

[45] Lipski, 23.
[46] Johnston, 11.

Current Medical Treatment for IBS and IBD

There are many available drugs for the treatment of both IBS and IBD. However, the use of these drugs also comes with side effects, sometimes worse than the disease itself. The most effective treatment would be to have minimal or no side effects, and EFA's appear to be a promising alternative.

At the time that this paper was originally written, Zelnorm was the newest treatment for constipation-predominant IBS in women. It binds to the 5-HT4 serotonin receptor and blocks it, preventing the serotonin from attaching to it. When serotonin binds to this receptor, it prevents intestinal contractions. So, by blocking the receptor, there is an increase in the contractions of the intestine which helps to prevent constipation. In addition to the side effects of headache and diarrhea, the following are the listed precautions according to the web site of Novartis, the company that manufactures this medication:

> "Zelnorm is contraindicated in those patients with severe renal impairment; moderate or severe hepatic impairment; a history of bowel obstruction, symptomatic gallbladder disease, suspected sphincter of Oddi dysfunction, or abdominal adhesions..."[47]

Since this paper was written, the FDA has taken Zelnorm off the market except for use in emergency cases. In March, 2007, the FDA recalled Zelnorm because of a link between the use of the drug and heart attacks, stroke, and angina. In July of the same year, the drug was listed as an "IND" or investigational new drug and it was made available only to patients who met certain criteria for its use. However, the manufacturer of Zelnorm discontinued this program in April, 2008, and the drug is now only available in emergency situations.

The most common medication used for diarrhea-predominant IBS is Loperamide. This medication slows the contractions of the intestinal wall and may or may not affect abdominal pain. This medication can cause constipation so it must be carefully monitored. Additional possible side effects include abdominal pain, dizziness, nausea, vomiting and dry mouth. Severe cases of diarrhea-predominant IBS which have not responded to other treatments may be treated with Alosetron (Lotronex). It also affects

---

[47] About Zelnorm 2004. Novartis Pharmaceuticals. 10 Jan 2004
<http://www.zelnorm.com/hcp/about/zelnorm/importsafety.jsp>

serotonin receptors as it blocks the 5-HT3 receptor resulting in a reduction in intestinal contractions and has been documented to have serious side effects.

Since this paper was originally written, Lotronex has been voluntarily recalled by its manufacturer after discussions with the FDA about serious side effects associated with the medication. The concern for the drug's safety emerged after several reports of intestinal damage from cases of ischemic colitis that occurred while patients were taking the drug. As of 2000, 70 adverse events had been reported to the FDA including ischemic colitis and severe constipation. Three of these cases resulted in death of the patient.

Crohn's disease is treated by a number of different medications. 5-aminosalicylic acid, or 5-ASA, has been used for relief of symptoms by putting patients in remission for longer periods of time than placebos. Two 5-ASA treatments, Asacol and Pentasa, both inhibit leukotriene synthesis that occurs from the conversion of AA in the lipoxygenase pathway. Asacol dissolves in the ileum and cecum, while Pentasa releases 5-ASA throughout the entire small intestine. The side effects from these drugs include headache, diarrhea, and skin rash, and they also decrease the effects of iron and folic acid in the body. The antibiotics metronidazole (Flagyl) and ciprofloxacin have been found to be useful although they also have possible side effects including:

Metronidazole - possible seizures and peripheral neuropathy, possibly carcinogenic with long term treatment.

Ciprofloxacin - possible adverse effects on renal function, super infection with long term treatment.

Corticosteroids are used if there is no response to 5-ASA and/or the antibiotics. Budesonide and Prednisone are corticosteroids that are commonly used, and they help to release anti-inflammatory cytokines as well as decrease the production of eicosanoids. Some side effects from the use of corticosteroids are as follows:

Hyperglycemia
Osteoporosis
Psychosis
Edema
Peptic ulcer disease
Infections

Immunosuppressive drugs have also been used, such as 6-mercaptopurine (6-MP) and methotrexate. They interfere with protein synthesis and nucleic acid metabolism. Possible side effects include:

Rash
Fever
Nausea
Vomiting
Diarrhea
Anemia
Hepatitis

Methotrexate has toxic effects on many bodily systems including the gastrointestinal tract, nervous system and circulatory system. Some side effects can be life threatening.

Treatments for ulcerative colitis include anti-inflammatory and/or immunosuppressive drugs. As described above, Asacol and Pentasa can be used. If these are ineffective, immunosuppressants can be used, but these drugs have many side effects (see above). Some other common drugs used in the treatment of ulcerative colitis along with their side effects include:

Azathioprine - loss of appetite, nausea, vomiting
Cyclosporine - shaky hands, headache, nausea, and leg cramps

If the above medical treatment for Crohn's disease fails, surgery may be considered, although this is not a cure. Re-occurrence of the disease is common. However, surgery can be a cure in ulcerative colitis, but this is an option taken only after all other treatment has failed.

The above medications have been shown to be of benefit in the treatment of these disorders. However, some of the medications have side effects that are probably worse than the disease itself! The most promising treatment in all of these digestive disorders is increasing the consumption of omega-3 fatty acids. As seen in the previously mentioned studies, omega-3's have been repeatedly shown to reduce inflammation, alter serotonin signaling, and reduce disease activity. In addition, these EFA's may be able to accomplish this without the horrible side effects of drugs. To prevent the patient from having to go through the side effects of the above drugs, it is a good idea to try and change one's diet and increase the intake of omega-3 fatty acids through the consumption of fish oil or flaxseed to see if this will eliminate or at least reduce the symptoms to a tolerable level.

World Demographics Comparing Rates of IBS and IBD

As stated previously, the rate of IBD has grown steadily in the United States since the 1930's. The rates are similar in Europe and Canada, but this disease is rare in South America, African blacks and Asia. In fact, researchers have found that IBD is predominately a disease of developed countries. More cases are reported in the northern hemisphere than the southern hemisphere. An exception to this is Australia, but it is important to realize that Australia is also a developed country. Those with a processed "western diet" report more cases than those that eat a more natural diet. In addition, more cases are reported in rural areas as opposed to urban areas. Jews from Eastern Europe are affected at a rate 4 to 5 times the general population. Caucasians and blacks in the U.S. are affected at about the same rate, and Asians and Hispanics are affected the least.

The interest in omega-3 fatty acids and their role in health came with the observation that Eskimos tend to have very low rates of heart disease and IBD[48], even though their diet was high in fat. However, the fat they were consuming was high in omega-3 fatty acids. Also, when they change to a North American diet, their heart disease risk increases to that seen in the United States. According to figures in The Omega Rx Zone by Dr. Barry Sears, the levels of DHA found in human breast milk were highest in the Eskimos in Canada at 1.4% DHA. Malaysia, China and Japan ranked 2nd, 3rd and 4th respectively. The United States came in 8th with only 0.23% DHA.[49]

Japan was once considered one of the healthiest countries as it had rare cases of heart disease and prostate, breast and colon cancers. IBD was also reported to be low in this country. Their traditional diet was low in saturated and hydrogenated fats and included high amounts of omega-3 fatty acids because of their high fish intake. In fact, their fish intake is much higher than in other Asian countries, and one of their staples is sashimi which is a type of raw fish.[50] Their diet was also high in fiber. However, Japan is more westernized than other Asian countries, and as they switch to a more modern day diet, they are developing more health problems. The Japan Hospital Association reported that the health of the Japanese people fell to a 15 year low in 1984.[51]

---

[48] A. Belluzzi et al., "Polyunsaturated Fatty Acids and Inflammatory Bowel Disease" <u>American Journal of Clinical Nutrition</u> 71 (1 Suppl) (Jan 2000). 26 Feb 2004
<http://www.ncbi.nlm.nih.gov/entrez/query.fcgi?cmd>

[49] Sears, 195.

[50] Elson M. Haas, <u>Staying Healthy With Nutrition: The Complete Guide to Diet and Nutritional Medicine</u> Celestial Arts. 4 Feb 2004 <http://www.healthy.net/asp/templates/article.asp?PageType=article&ID=1700>

In addition, low rates of IBD have been noted in China. In the rural areas of China, the diet is an overall healthy one with rice being a staple. Very small amounts of animal food are eaten, but vegetables constitute a large part of the diet. The concern about health issues in this country is in the city where the availability of processed and junk foods is evident.[52]

In southern Africa, the diet is more "westernized" and is similar to what is seen in the U.S. and England. However, the black Africans tend to eat a more natural diet of vegetables, fish and grains. The rate of IBD in African blacks is low, and this natural diet could be the reason why these people are healthier.

The Mediterranean diet has been talked about by many for its health benefits. Although the rate of IBD appears to be lower in this region, more studies need to be done. This diet is very high in olive oil which contains the omega-9 oleic acid. In addition, fish and other seafood are abundant in this diet along with fresh fruits and vegetables. The Mediterranean diet has more total fat overall than the diet of northern Europe, but the fat is predominately unsaturated. The diet of those living in Crete consists of 40 percent total fat with the majority being unsaturated.[53] This diet also has a 1-2 to 1 ratio of omega-6 to omega-3 fatty acids and, as previously discussed, this is a much more healthy ratio of these essential fats than what is seen the western diet.[54]

Scandinavia is an interesting area as well because of their high reported rate of IBD. Although the diet does include lots of fish, the country has become more "westernized" and has increased availability of processed foods.[55] This increase in food processed products leads to an increase in omega-6 fatty acids in the diet which may lead to potential health problems.

Eurpoean Jews also have a very high occurrence rate of IBD. They ten to use more flour products rather than whole grains, and they tend to eat fruit cooked. Each meal usually contains one animal protein, and

---

[51] Weil, 158.
[52] Haas.
[53] Artemis P. Simopoulos, M.D. <u>Evolutionary Aspects of Diet and Essential Fatty Acids</u> 31 Oct 2000. Kronos Institute Seminar Series. 1 Mar 2004 <http://kronosinstitute.org/seminars/seminar-2000-10-31-simopoulos.html>
[54] Simopoulos.
[55] <u>Welcome to America</u> 2003. Health Day News. 4 Feb 2004 <http://www.hon.ch/News/HSN/51345.html>

consumption of whitefish is common.[56] Their diet consists of a lot of processed food and less natural food such as fresh fruits and vegetables. In fact, Israeli Jews have one of the highest rates of omega-6 consumption in the world.[57] This may explain the high rate of IBD seen in this culture.

A study by Sood et al. in 2003 showed some interesting results on the incidence of ulcerative colitis in Punjab, North India. The conclusion of the researchers stated that the "…disease frequency is not much less than that reported from Europe and North America."[58] However, if we look at the traditional Indian diet, fish consumption is not high, and very few raw foods are eaten. Also, they tend to eat a fair amount of fried foods.

It is difficult to obtain accurate epidemiological data for IBS because studies have been based on self-reporting of IBS instead of using the standard diagnostic criteria. Hungin et al. determined through their study that the occurrence rate for this disorder is about 14 percent and that women are affected twice as often as men.[59] Similar figures have been seen in Canada and Europe.

Additional studies need to be done on the diets of other countries and their rated of IBD and IBS. Although not completely clear, there does appear to be a link between the imbalance of omega-6 to omega-3 fatty acids and the development of these inflammatory conditions. The levels of these essential fatty acids need to be measured since some diets, although high in fish consumption, are also high in processed foods. Overall, it appears that the "westernized" diet promotes the development of these disorders probably through the destruction of omega-3 and the dramatic increase in omega-6 fatty acids.

---

[56] Haas.
[57] Natural Healing with Flax: Important Help for Diabetics 2003. Barlean's Organic Oils, L.L.C. 1 Mar 2004 <http://www.barleans.com/literature/flax/83-help-for-diabetics.html>
[58] A. Sood et al., "Incidence and Prevalence of Ulcerative Colitis in Punjab, North India" Gut Volume 52 (2003) 5 Feb 2004 <http://gut.bmjjournals.com/cgi/content/abstract/52/11/1587>
[59] Nicholas J. Talley, New and Important Insights into IBS: From Epidemiology to Treatment 30 Jan 2003. DreamPharm.com 4 Feb 2004 <http://dreampharm.com/he;atitis-info2/irritable-bowel-syndrome-1-30ibs-03.html>

Conflicting Views and Medical Studies

Most studies found seemed to suggest a positive benefit from omega-3 fatty acid treatment of both IBS and IBD. However, there are a few studies that do not support this view. The following studies found inconsistent results with the previous studies mentioned.

In 1993, Greenfield et al. performed a 6 month, placebo-controlled study testing 43 patients with ulcerative colitis. The patients were divided into three groups and received either MaxEPA, super evening primrose oil or olive oil. Although red cell membrane concentrations of omega-3 fatty acids were increased in the groups receiving MaxEPA and DGLA concentrations increased in those receiving evening primrose oil, there was no noted improvement in disease relapse, rectal bleeding, stool frequency or appearance of the intestinal wall through sigmoidoscopy. They concluded that "Despite manipulation of cell-membrane fatty acids, fish oils do not exert a therapeutic effect in ulcerative colitis, while evening primrose oil may be of some benefit."[60]

In a study performed by Hawthorne et al. and published in Gut in 1992, 87 patients were studied over the course of a year. They were divided into two groups. One group received 20 ml. HiEPA fish oil, and the other group received olive oil as a placebo. The EPA content of the rectal mucosa of those taking the fish oil rose to 3.2 percent of total fatty acids, whereas the EPA content was only 0.63 percent total fatty acids for those on the placebo. There was an increase in leukotriene B5 in addition to a 53 percent decrease in leukotriene B4 synthesis. Patients on the fish oil tended to go into remission faster than those on the olive oil, although the researchers felt the differences were insignificant. They concluded that fish oil supplementation was of limited benefit with a modest effect in reduction of corticosteroid use.[61]

---

[60] S.M. Greenfield et al., "A Randomized Controlled Study of Evening Primrose Oil and Fish Oil in Ulcerative Colitis" Alimentary Pharmacological Therapy 7/2 (1993). 21 May 2002
<http://www.lef.org/prod_hp/abstracts/php-ab161b.html>

[61] A.B. Hawthorne et al., "Treatment of Ulcerative Colitis with Fish Oil Supplementation: A Prospective 12 Month Randomised Controlled Trial" Gut 33/7 (1992). 21 May 2002
<http://www.lef.org/prod_hp/abstracts/php-ab161b.html>

Conclusion

According to Abraham Hoffer, M.D., PhD, "If the patient has been to more than four physicians, nutrition is probably the medical answer."[62]

Irritable bowel syndrome and inflammatory bowel disease are two disorders of the gastrointestinal tract that can seriously disrupt the lives of the affected patients. An effective form of treatment is necessary to help relieve the discomfort that these patients experience on a daily basis.

It is important to understand how much of an impact diet can have on health. In 1986, I suffered from a ruptured appendix, and the pain and discomfort was excruciating. However, IBS, resulting from a poor diet and lack of omega-3 fatty acids, resulted in greater pain and discomfort than a ruptured appendix! I would have never guessed that diet would affect my health to that degree, and I believe that most people probably feel the same way. So, it is crucial to understand that a poor diet can cause tremendous pain and discomfort and can seriously disrupt a person's life.

IBS is a disorder of the functioning of the gastrointestinal tract. It appears that this malfunctioning could be due to abnormal serotonin signaling in the intestinal wall. The omega-3 fatty acids have been documented to alter serotonin levels and may be effective in alleviating the symptoms. These fatty acids, whether obtained through diet, fish oil or flax oil appear to be an effective treatment and do not have the side effects of medication. Additionally, these fatty acids have other benefits such as promoting heart health by lowering cholesterol and reducing blood pressure, improving symptoms of autoimmune diseases such as rheumatoid arthritis, and alleviating symptoms of depression.

IBD is a chronic condition affecting the digestive tract. Inflammation seems to be a predominant feature and may be linked to the production of pro-inflammatory cytokines as a result of eating a diet high in omega-6 fatty acids. Studies involving the use of omega-3 fatty acids in the treatment of IBD have shown promising results through their anti-inflammatory effects. In addition to physicians and gastroenterologists, other experts need to be involved in the treatment of IBD, including psychologists and nutritionists.

---

[62] Lipski, 3.

In new and mild cases, nutrition is probably the best place to start in treatment as this may give relief without the side effects of medication. In more advanced cases, medication or intravenous feeding may be necessary because of the inability of the intestine to absorb nutrients due to inflammation; however, proper nutrition should be a factor in any treatment program. Continued poor food choices will only exacerbate the condition and will lead to further health problems, even with medication. In addition, the success of the treatment depends on the willingness and cooperation of the patient. If the patient refuses to work with the physicians and nutritionists, his/her health will continue to suffer. As explained previously, changing only one factor is futile. The patient must be willing to make lifestyle and diet changes and must be willing to work with all involved.

In order to help reduce the number of these and other inflammatory disorders, there needs to be a push toward education to help people to understand why these inflammatory diseases are on the rise. One way to accomplish this is to include a mandatory nutrition course in both elementary and high schools. This course should focus purely on nutrition and its role in health. In addition, medical schools should be offering nutrition courses, and it should be mandatory for students to take several of these courses prior to graduation. According to the Rockefeller Institute of Medical Research,

"If the doctors of today do not become the nutritionists of tomorrow, then the nutritionists of today will become the doctors of tomorrow."[63]

With nutrition knowledge, which appears to be lacking in today's medical field, better overall treatment can be offered to those suffering from these kinds of disorders. I can comment on this from personal experience as I was not given any nutritional advice from any doctor during the years that I was under medical treatment for IBS. Yet, when I changed my diet and started supplementation with omega-3 fatty acids, the disorder improved dramatically. Mandatory nutrition classes in schools will give people more power in controlling their health and possibly help prevent some of these conditions before they start. This could potentially decrease medical costs in the future.

---

[63] Fats and Oils Live Renewed Life.com 10 Feb 2001 <http://www.liverenewedlife.com/fats_oils.htm>

Through education and additional research studies, the causes and cures to these digestive disorders will become clearer. It is critical that we find the cause and proper treatment for these disorders so we can prevent future suffering. Omega-3 fatty acids have been shown to help in both IBS and IBD and should be a part of the treatment program.

Bibliography

About Zelnorm. 2004. Novartis Phamaceuticals. 10 Jan 2004
    <http://www.zelnorm.com/hcp/about/zelnorm/importsafety.jsp>

Asian, A. and G. Triadafilopoulos. "Fish Oil Fatty Acid Supplementation in
    Active Ulcerative Colitis: A Couble Blind, Placebo-Controlled, Crossover Study" American
    Journal of Gastroenterology. 87/4 (1992) 21 May 2002
    <http://www.lef.org/prod_hp/abstracts/php- ab161b.html>

Belluzzi, A. et al. "Polyunsaturated Fatty Acids and Inflammatory Bowel Disease" American Journal of
    Clinical Nutrition. 71 (Suppl 1) (Jan 200). 26 Feb 2004
    <http://www.ncbi.nlm.nih.gov/entrez/query.fcgi?cmd=Retrieve&db=
    PubMed&list_uids=10617993&dopt=Abstract>

Belluzzi, Andrea et al. "Effect of an Enteric-Coated Fish-Oil Preparation on Relapses in Crohn's Disease"
    The New England Journal of Medicine. Volume 334, Number 24 (13 Jun 1996). 23 Feb 2004
    <http://content.nejm.org/cgi/content/abstract>

Christensen, James M.D. and Robert Summers, M.D. Understanding IBS (Irritable Bowel Syndrome). July
    2003. The University of Iowa: Virtual Hospital. 10 Feb 2004
    <http://www.vh.org/adult/patient/internalmedicine/irritablebowel
    syndrome/index.html>

Davis, Bowman O. "Gastrointestinal Pathophysiology". 2000. Kennesaw State University. 24 Jan 2004
    <http://science.kennesaw.edu/~bodavis/LECT13GI.htm>

The Enteric Nervous System. 23 Aug 1998. Colorado State University. 22 Nov 2003
    <http://arbl.cvmbs.colostate.edu/hbooks/pathphys/digestion/basics/gi_nervous.html>

Fats and Oils. Live Renewed Life.Com 10 Feb 2004 <http://www.liverenewedlife.com/fats_oils.htm>

Flax Oil May Aid Persons with Crohn's Disease. 2003. Barleans OrganicOils, L.L.C. 16 Dec 2003
    <http://www.barleans.com/literature/flax/116-flax-and-crohns.html>

Ganong, William. Review of Medical Physiology. San Fransisco: McGraw Hill, 2003.

Gilston, V. "Inflammatory Mediators, Free Radicals and Gene Transcription" Progress in Inflammation
    Research. (1999) 28 Feb 2004 <http://www.birkhauser.ch/books/biosc/pir/pir5851toc.html>

Goldberg, Burton. Alternative Medicine, The Definitive Guide. Tiburon: Future Medicine Publishing, Inc.,
    1999.

Greenfield, S.M. et al. "A Randomized Controlled Study of Evening Primrose Oil and Fish Oil in
    Ulcerative Colitis" Alimentary Pharmacological Therapy. 7/2 (1993). 21 May 2002
    <http://www.lef.org/prod_hp/abstracts/php-ab161b.html>

Grimminger, F. et al. "Influence of Intravenous n-3 Lipid Supplementation of Fatty Acid Profiles and Lipid
    Mediator Generation in a Patient with Severe Ulcerative Colitis" European Journal of Clinical
    Investigation. 23(1) (1993). 21 May 2002 <http://www.lef.org/prod_hp/abstracts/php-b161b.html>

Haas, Elson M. Staying Healthy With Nutrition: The Complete Guide to Diet and Nutritional Medicine. Celestial Arts. 4 Feb 2004 <http://www.healthy.net/asp/templates/article.asp?PageType=article&ID=1700>

Hawthorne, A.B. et al. "Treatment of Ulcerative Colitis with Fish Oil Supplementation: A Prospective 12 Month Randomised Controlled Trial" Gut 33/7 (1992). 21 May 2002 <http://www.lef.org/prod_hp/abstracts/php-ab161b.html>

Hillier, K. et al. "Incorporation of Fatty Acids from Fish Oil and Olive Oil Into Colonic Mucosal Lipids and Effects Upon Eicosanoid Synthesis in Inflammatory Bowel Disease" Gut 32/10 (1991). 21 May 2002 <http://www.lef.org/prod_hp/abstracts/php-ab161b.html>

IBD and Fatty Acids. 2002. Great Smokies Diagnostic Laboratory. 21 May 2002 <http://www.gsdl.com/assessments/finddisease/ibd/fatty_acids.html>

Inflammatory Bowel Disease. 2002. UCSF Gastroenterology Division. 14 Jan 2004 <http://gi.ucsf.edu/basicIBD.html>

Irritable Bowel Syndrome. 16 June 2003. MedicineNet, Inc. 10 Jan 2004 <http://www.medicinenet.com/Irritable_Bowel_Syndrome/page5.htm>

Johnston, Ingeborg M. and James R. Johnston, PhD. Flaxseed (Linseed) Oil and the Power of Omega-3. Los Angeles: Keats Publishing, 1990.

King, Michael. Biochemistry of Neurotransmitters. 12 Aug 2003. IU School of Medicine. 10 Jan 2004 <http://www.indstate.edu/thcme/mwking/nerves.html>

Kucharzik, Torsten et al. "Synergistic Effect of Immunoregulatory Cytokines on Peripheral Blood Monocytes from Patients with Inflammatory Bowel Disease" Digestive Diseases and Sciences 42(4) (1 Apr 1997). 14 Jan 2004 <http://www.ncbi.nlm.nih.gov/entrez/query.fcgi?cmd=Retrieve&db=PubMed&list_uids=9125653&dopt=Abstract>

Lipski, Elizabeth. Digestive Wellness. Los Angeles: Keats Publishing, 2000.

Li, H. D. Liu and E Zhang. "Effect of Fish Oil Supplementation on Fatty Acid Composition and Neurotransmitters of Growing Rats" Wei Sheng Yan Jiu 29(1) (30 Jan 2000). 14 Jan 2004 <http://www.ncbi.nlm.nih.gov/entrez/query.fcgi?cmd=Retrieve&db=PubMed&list_uids=12725043&dopt=Abstract>

Mamula, Petar and Jonathan Markowitz. "Crohn Disease" eMedicine Journal Volume 3, Number 1 (16 Jan 2002). 15 Jun 2002 <http://www.emedicine.com/ped/topic507.htm>

Murray, Michael N.D. Boost Your Serotonin levels: 5-HTP, The Natural Way to Overcome Depression, Obesity and Insomnia. New York: Bantam Books, 1998

Nachbur, Jennifer. UVM Researchers Identify Molecular Changes in IBS Patients. 22 Oct. 2003. The University of Vermont. 10 Jan 2004 <http://www.uvm.edu/news/print/?action=Print&StoryID=4188>

Natural Healing with Flax: Important Help for Diabetics. 2003. Barlean's Organic Oils, L.L.C. 1 Mar 2004 <http://www.barleans/com/literature/flax/83-help-for-diabetics.html>

Nicol, Rosemary. Irritable Bowel Syndrome: A Natural Approach. Berkeley: Ulysses Press, 1999

Null, Gary. Healing Your Body Naturally: Alternative Treatments to Illness. New York: Seven Stories, 1997

Omega-6 Fatty Acids. 2001. University of Maryland Medicine. 26 Feb 2003
<http://www.umm.edu/altmed/ConsSupplements/Omega6FattyAcidscs.html>

Pakala, R. et al. "Eicosapentaenoic Acid and Docosahexaenoic Acid Block Serotonin Induced Smooth Muscle Cell Proliferation" Arteriosclerosis, Thrombosis and Vascular Biology 19(10 (Oct 1999). 14 Jan 2004
<http://www.ncbi.nlm.nih.gov/entrez/query.fcgi?cmd_Retrieve&db=PubMed&list_uids=10521359&dopt=Abstract>

Pathophysiology of IBS. 2004. Novartis Pharmaceuticals Corporation. 10 Jan 2004
<http://www.zelnorm.com/hcp/about/ibs/patho.jsp?checked=y>

Peikin, Steven. Gastrointestinal Health. New York: Harper Collins, 1991

Pressman, Alan H. and Sheila Buff. The Complete Idiots Guide to Vitamins and Minerals. Indianapolis: Alpha Books, 2000

"Prevalence of Essential Fatty Acid Deficiency in Patients with Chronic Gastrointestinal Disorders" Metabolism. 45(1)(Jan 1996). 21 May 2002 <http://www.lef.org/prod_hp/abstracts/php-ab161b.html>

Rank, Angela. Eicosanoids. Wake Forest University. 10 Jan 2004
<http://www.wfu.edu/users/clafme0/nutrition/eicosanoids.htm>

Reif, S. et al. "Pre-illness Dietary Factors in Inflammatory Bowel Disease" Gut Vol 40 (1997). 26 May 2002 <http://gut.bmjjournals.com/cgi/content/abstract/40/6/754>

Ross, E. "The Role of Marine Fish Oils in the Treatment of Ulcerative Colitis" Nutrition Review 51/2 (1993). 21 May 2002 <http://www.lef.org/prod_hp/abstracts/php-ab161b.html>

Rudin, Donald and Clara Felix. Omega-3 Oils, A Practical Guide. U.S.: Avery, 1996.

Saibil, Fred. Crohn's Disease and Ulcerative Colitis, Everything You Need to Know. New York: Firefly, 1996.

Sears, Barry. The Omega Rx Zone: The Miracle of the New High-Dose Fish Oil. New York: Harper Collins Publishers, Inc., 2002.

Serotonin and IBS. 19 Jan 2003. International Foundation for Functional Gastrointestinal Disorders, Inc. 10 Jan 2004 <http://www.aboutibs.org/Publications/serotonin.html>

Simopoulos, A.P. "Omega-3 Fatty Acids in Inflammation and Autoimmune Diseases". Journal of the American College of Nutrition 21(6) (Dec 2002). 16 Dec 2003
<http://www.ncbi.nlm.nih.gov/entrez/query.fcgi?cmd=Retrieve&db=PubMed&list_uids=12480795&dopt=Abstract>

Simopoulos, Artemis P. Evolutionary Aspects of Diet and Essential Fatty Acids. 31 Oct 2000. Kronos Institute Seminar Series. 1 Mar 2004 <http://kronosinstitute.org/seminars/seminar-2000-10-31-simopoulos.html>

Sood, A. et al. "Incidence and Prevalence of Ulcerative Colitis in Punjab, North India". Gut Volume 52 (2003). 5 Feb 2004 <http://gut.bmjjournals.com/cgi/content/abstract/52/11/1587>

Talley, Nicholas J. New and Important Insights into IBS: From Epidemiology to Treatment. 30 Jan 2003. DreamPharm.com. 4 Feb 2004 <http://dreampharm.com/hepatitis-info2/irritable-bowel-syndrome-1-30ibs-03.html>

Tagore, A. et al. "Interleukin-10 (IL-10) Genotypes in Inflammatory Bowel Disease" Tissue Antigens Volume 54, Issue 4 (Oct 1999). 14 Jan 2004 <http://www.blackwellenergy.com/servlet/useragent?fun=synergy&synergyAction=showAbstract&doi>

Wang, Q.Y. et al. "Expression of Pro-Inflammatory Cytokines and Activation of Nuclear Factor KappaB in the Intestinal Mucosa of Mice with Ulcerative Colitis" Di Yi Jun Yi Da Xue Xue Bao 23(11) (Nov 2003). 14 Jan 2004 <http://www.ncbi.nlm.nih.gov/entrez/query.fcgi?cmd=Retrieve&db=PubMed&list_uids=14625189&dopt+Abstract>

Weil, Andrew. Eating Well For Optimum Health. New York: Alfred A. Knopf, 2000.

Welcome to America. 2003. Health Day News. 4 Feb 2004 <http://www.hon.ch/News/HSN/513545/html>